Azure Internet of Things Revealed

Architecture and Fundamentals

Robert Stackowiak

Apress®

Azure Internet of Things Revealed: Architecture and Fundamentals

Robert Stackowiak
Elgin, IL, USA

ISBN-13 (pbk): 978-1-4842-5469-1   ISBN-13 (electronic): 978-1-4842-5470-7
https://doi.org/10.1007/978-1-4842-5470-7

Managing Director, Apress Media LLC: Welmoed Spahr
Acquisitions Editor: Jonathan Gennick
Development Editor: Laura Berendson
Coordinating Editor: Jill Balzano

Cover image designed by Freepik (www.freepik.com)

Distributed to the book trade worldwide by Springer Science+Business Media New York, 233 Spring Street, 6th Floor, New York, NY 10013. Phone 1-800-SPRINGER, fax (201) 348-4505, e-mail orders-ny@springer-sbm.com, or visit www.springeronline.com. Apress Media, LLC is a California LLC and the sole member (owner) is Springer Science + Business Media Finance Inc (SSBM Finance Inc). SSBM Finance Inc is a **Delaware** corporation.

For information on translations, please e-mail rights@apress.com, or visit http://www.apress.com/rights-permissions.

Apress titles may be purchased in bulk for academic, corporate, or promotional use. eBook versions and licenses are also available for most titles. For more information, reference our Print and eBook Bulk Sales web page at http://www.apress.com/bulk-sales.

Any source code or other supplementary material referenced by the author in this book is available to readers on GitHub via the book's product page, located at www.apress.com/9781484254691. For more detailed information, please visit http://www.apress.com/source-code.

Printed on acid-free paper

*Dedicated to my wife and long-time partner Jodie
–the adventure continues*

Table of Contents

About the Author xi

Acknowledgments xiii

Introduction xv

Chapter 1: Modern IoT Architecture Patterns 1

  The Evolution of the Internet of Things 2

  Typical IoT-Based Business Solutions 4

     Agribusiness Examples 5

     Automotive Examples 5

     Aviation Examples 6

     Communications and Media Transmission Examples 6

     Construction Examples 7

     Consumer Packaged Goods Examples 7

     Education and Research Examples 8

     Environmental Controls Examples 8

     Financial Banking and Trading Firm Examples 9

     Healthcare Payers and Providers Examples 9

     High Tech and Industrial Manufacturing Examples 10

     Insurance Company Examples 11

     Law Enforcement and Emergency Services Examples 11

     Media Content and Entertainment Examples 11

     Oil and Gas Examples 12

     Pharmaceutical and Medical Device Examples 12

     Retail Examples 13

     Transportation and Logistics Examples 13

     Utility Company Examples 14

IoT Reference Architectures..15

How IoT Fits in Your IT Architecture 17

Why Cloud Computing and IoT 21

Other IoT Concepts and Considerations 24

An Evolution in Needed Skills 26

Chapter 2: Azure IoT Solutions Overview 29

Microsoft Azure PaaS and IoT 30

    Azure IoT Hub 32

    Azure Digital Twins 33

    Azure Stream Analytics 33

    Azure Time Series Insights 34

    Azure Databricks 34

    Azure Data Lake Storage 35

    Azure HDInsight..35

    Cosmos DB 37

    Other Azure Data Stores 37

    Tools, Frameworks, and Services...37

Non-Microsoft Components in Azure IoT 38

IoT SaaS Solutions in Azure 40

Azure Management and Deployment...41

    Subscriptions and Resource Groups 41

    Azure Portal..43

    Designing for Resiliency and Availability 47

    Azure Security Considerations...50

Microsoft Intelligent Edge 51

    Azure IoT Edge..52

    Azure Sphere 52

    Windows 10 IoT 53

Choosing the Right Component Model...54

Chapter 3: IoT Edge Devices and Microsoft 55

Edge Sensor and Device Selection 56

The Azure IoT Edge Runtime 61

    The IoT Edge Device As a Gateway Device 62

    Deployment of Containers 64

    Azure IoT Edge and Device Security 65

The Azure IoT Device Catalog 68

Chapter 4: Azure IoT Hub 73

IoT Hub Capabilities 74

Configuring the IoT Hub 75

    Managing the IoT Hub 78

    Message Routing and Event Routing 79

IoT Hub Performance Monitoring 81

IoT Hub Device Provisioning 83

IoT Hub Availability and Disaster Recovery 84

Chapter 5: Analyzing and Visualizing Data in Azure 87

Azure Stream Analytics 88

Time Series Insights 89

Azure Databricks 91

Semi-structured Data Management 95

    Azure HDInsight 95

    Cosmos DB 97

Azure Machine Learning 99

    Azure Machine Learning Studio 99

    Azure Machine Learning Service 101

Cognitive Services 104

Data Visualization and Power BI 108

Azure Bot Service and Bot Framework 117

Chapter 6: IoT Central and Solution Accelerators 119

Azure IoT Central 120

IoT Solution Accelerators 129

Remote Monitoring 132

Predictive Maintenance 134

Connected Factory 138

Device Simulation 141

Chapter 7: Infrastructure Integration 145

Preexisting Sources of Data 145

Integrating and Finding Data Sources 147

Azure Data Factory 148

Query Services Across Diverse Data 151

Connecting On-Premises Networks to Azure 153

Bulk Data Transfer 156

Azure Data Catalog 158

Data Historians and Integration to Azure 160

Chapter 8: Developing a Plan for Success 163

Identifying the Right Initiatives 164

Observe and Research 167

Problem Definition 169

Ideation 173

Prototype Creation 180

Testing 181

The Agile Sprint Approach 183

Moving from Prototypes to Implementation 184

Measurable Return on Investment 185

Operational Considerations 187

Implementation Strategy 189

Preparing an Implementation Roadmap 190

Some Final Thoughts 191

Appendix: Published Sources **193**

Microsoft Online Documentation Sources.. 194

Other Web Site Sources 196

Index.. **199**

About the Author

 **Robert Stackowiak** is Data and Artificial Intelligence Architect and Technology Business Strategist at the Microsoft Technology Center in Chicago, Illinois, USA. He regularly conducts executive briefings, business discovery workshops, and technology architecture sessions with many of North America's most leading-edge companies across a variety of industries and with government agencies. Bob has spoken at numerous industry conferences internationally, served as a guest instructor at various universities, and is an author of several books, including *Remaining Relevant in Your Tech Career* (Apress), *Big Data and the Internet of Things* (Apress), *Oracle Big Data Handbook,* and *Oracle Essentials.* He joined Microsoft in May 2016 after a 20-year career at Oracle where he was most recently Executive Director of Big Data. You can follow him on Twitter @rstackow, and read his articles and posts on LinkedIn.

Acknowledgments

I have learned a lot about the Internet of Things since being the lead author on the topic for an Apress book about 4 years ago. My clients have deployed a variety of these solutions and increasingly look for cloud-based resources, such as those found in Microsoft Azure, to scale the analysis of incoming data. More recently, many of these clients are also deploying Microsoft's intelligent edge.

I have found particularly useful my meetings with major manufacturers and other companies, especially as they shared their vision and experiences. Many are using data from sensors for monitoring their connected factories, monitoring the remote devices that they build, and analyzing data from equipment and devices in performing predictive maintenance. Thanks to them for sharing.

During IoT briefings for clients, I frequently collaborated with other specialists within Microsoft. I would like to call out Michael Walton and Eddy Saad for helping our clients see a vision of what is possible.

Microsoft has a broad array of partners helping clients design and deploy these solutions. Some of those that I have worked with include Accenture, Avanade, and Hitachi Solutions, as well as many of Microsoft's analytics partners and solution providers. Thanks especially to Jerry Hawk and his team for their support and guidance in many of these opportunities.

As I wrote this book, extensive changes were occurring in the Chicago Microsoft Technology Center where I worked. I would especially like to thank my former Director, Beth Malloy, and my current Director, Adetayo Adegoke, and long-time associates Charles Drayton and Ross LoForte for their support.

Of course, this book would not be possible without the efforts of Microsoft engineers and product managers who bring these offerings to market. The documentation that they produced was extremely useful in building the content presented here. Thanks for continuing to respond to our clients' needs by providing a comprehensive set of offerings.

Writing and publishing a book usually takes about 9 months, and that period can be a time of smooth collaboration with a publisher (including continuity of support from a few key players) or one in which the key players come and go. I have experienced both.

ACKNOWLEDGMENTS

This is the third book that I have written for Apress, and they have become my favorite publisher. Jonathan Gennick, Assistant Editorial Director, once again saw the need and value for a book like this and gained swift approval for its production. He and I have now collaborated on producing books for over 20 years. Jill Balzano was once again the Coordinating Editor, my third collaboration with her. She makes staging a book like this one extremely easy.

This book was written on my long daily commutes on the train or evenings and weekends. It did require that I make some hard choices regarding where I would spend my time. Hopefully, my wife Jodie found that my writing of this book was not quite so noticeable as some of my earlier efforts. (Writers can get cranky at times.) The support from Jodie has been amazing over the years. Now that retirement from full-time work is just around the corner, I hope she doesn't mind if I write another book or two in the future!

Introduction

The Internet of Things (IoT) has been widely discussed and written about over the past decade. So, why did we believe this book was needed?

As many companies began deploying IoT solutions, new needs became evident. These needs drove advances in components and changes in IoT architecture. Microsoft was at the forefront of this development, responding to these requirements with new Azure backend analytics tools and data management solutions, an Azure IoT Hub enabling the landing data streams in the cloud and managing of devices, and pushing analytics and other functions to the devices themselves.

As the number of components and options for deployment has increased, Microsoft has sought to simplify deployment for certain scenarios by producing solution accelerators. Many of Microsoft's partners have also embraced some or all these components in commercial product and service offerings.

It occurred to us that the time was right to describe all these components and offerings in a single volume since the complexity of all of this can appear to be overwhelming. Our goal in this book is to explain the capabilities, options, and architecture patterns that you might choose to incorporate in your own designs and implementations.

The book begins with a chapter covering generic architecture patterns and key components in IoT. In the second chapter, we provide an overview of major Microsoft components typically found in the architecture.

In Chapters 3 through 7, we provide more detail regarding the Microsoft components and include chapters on IoT devices, the Azure IoT Hub, analyzing and visualizing data in Azure, IoT Central and the solution accelerators, and infrastructure integration considerations.

We conclude the book with a chapter describing how you might develop a plan for success using proven techniques that include design thinking.

We realize that this book is likely the start of your exploration of this topic or that you are using the book to refresh your knowledge with what is new and current. As further exploration of these topics will likely be desired, we've included an extensive list of sources at the end of the book.

We hope that you will find the book to be a valuable reference wherever you are on your IoT journey.

CHAPTER 1

Modern IoT Architecture Patterns

Today, Microsoft Azure footprints are often designed to be part of a broader architecture that includes Internet of Things (IoT) devices. Though you might be new to this type of solution, the need for such an architecture did not suddenly appear overnight. IoT itself has a long history that predates the cloud and Big Data.

Today's architectures feature highly scalable event handling enabling real-time analysis in what Microsoft has named the "intelligent cloud" and deployment of machine learning at the "intelligent edge" in the devices. As more advanced IoT solution components and capabilities have become available, previous architecture patterns evolved to take advantage of these new capabilities and enable more sophisticated business solutions to be deployed.

This chapter introduces IoT and covers its history and relevancy in solving a host of business problems in a variety of industries. We explain some of the basic terminology and typical architecture patterns that you will encounter. You should come away from this chapter ready to understand how Microsoft's technology components align to these patterns as we introduce them and then dig deeper into them throughout much of the remainder of the book.

Appropriately, this chapter is divided into these sections:

- The evolution of the Internet of Things

- Typical IoT-based business solutions

- IoT reference architectures

- How IoT fits in your IT architecture

© Robert Stackowiak 2019
R. Stackowiak, *Azure Internet of Things Revealed*, https://doi.org/10.1007/978-1-4842-5470-7_1

- Why cloud computing and IoT

- Other IoT concepts and considerations

- An evolution in needed skills

The Evolution of the Internet of Things

The Internet of Things (IoT) consists of sensors, devices, and/or actuators that are networked in order to gather data for processing and trigger actions or alerts enabling appropriate responses to be made. IoT architecture solutions are frequently deployed to enable intelligent and automated equipment that is deployed in homes, businesses, factories, vehicles, and outdoor locations. The products and solutions are designed to help solve industry specific problems and needs.

Intelligent devices at the edge of the architecture can both transmit and respond to data, sometimes by controlling other components or equipment present. Networking of the devices enables data sharing among them and transmission of data to a data center through a gateway for further processing and analysis. Today's IoT footprint can respond in real time and perform analysis on massive numbers of incoming events. This footprint represents the latest stage in the evolution of the key components in IoT.

The first device that many define as a sensor was the thermostat, invented in 1883. Motion sensors and infrared sensors first began to appear in the 1940s and the early 1950s. In the 1960s, sensors and associated computing devices were greatly reduced in size to meet the demands of the space program and were key in the development of spacecraft capable of landing men on the moon.

Networking software began to appear during this same time period to be used in linking computers and devices. The ARPANET was introduced in 1969 to transmit messages from computers and devices across wide distances, and it eventually evolved into the Internet. Early adopters of these networks included the oil and gas companies that needed to transmit exploration data gathered from sensors in drilling equipment to powerful backend computers used in performing analytics on the data.

RFID tags and UPC codes began to appear in the early 1970s, and widespread usage occurred in the following decade. By the late 1990s, RFID tags were linked to the Internet at MIT. Kevin Ashton referred to this work in a 1999 speech at Procter & Gamble as the "Internet of Things."

This was an era in which relational databases were commonly used to store and analyze all data. Data historians built upon relational database management systems became popular for analyzing time series data coming from sensors, programmable logic controllers (PLCs), and other similar devices.

In the early 2000s, new alternatives to relational databases began to gain wider adoption. Companies that built Internet search engines found that the data they needed for analysis arrived in streams and contained delimiters and other miscellaneous data intermixed with the data of value. The data streams required pre-processing to fit into relational databases since relational databases store data in tables neatly formatted into rows and columns. This data conversion introduced latency and complexity that soon became unacceptable to the search engine companies.

New database management systems were introduced to handle such semi-structured data streams. Often referred to as NoSQL databases, Hadoop clusters became especially popular initially for rapidly loading and analyzing large amounts of semi-structured data. Since data coming from many of the devices at the edge also was generated in a semi-structured form, IoT architectures began to include these new data management engines in the backend infrastructure. A "Lambda architecture," described in a subsequent section of this chapter, became popular in IoT deployment for handling streaming data and traditional batch data feeds.

Sensors continued to evolve, becoming smaller and cheaper, requiring less energy, and providing more functionality. The number of sensors and intelligent devices deployed experienced explosive growth throughout the 2010s.

New IoT use cases and growing data volumes drove a need to apply analytics and machine learning in real time at the location where the data was being gathered. Microsoft was among the first to refer to the devices containing sensors and featuring local compute capabilities as the intelligent edge.

Figure 1-1 illustrates the timeline of IoT evolution that we just described.

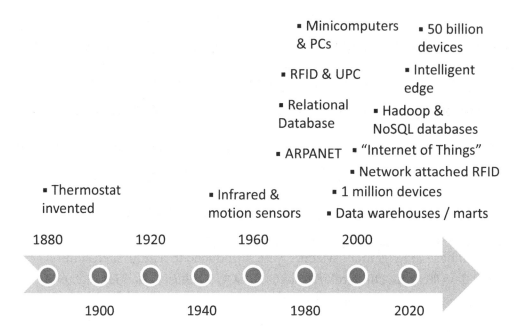

Figure 1-1. *Timeline of IoT evolution*

Before we look at how these technologies come together to form modern IoT architecture patterns, let's look at some of the IoT business solutions that leverage these patterns.

Typical IoT-Based Business Solutions

IoT architectures are used to solve a variety of business problems. The types of problems solved are often industry-dependent. Just as form follows function in classic architecture, one should first understand the kinds of problems that IoT solutions can solve and relevant business problems present in your company or organization before pursuing an IoT project.

In this section, we provide examples in agribusiness, automotive, aviation, communications and media transmission, construction, consumer packaged goods, education and research, environmental controls, financial banking and trading, healthcare payers and providers, high-tech and industrial manufacturing, insurance, law enforcement and emergency services, media content and entertainment, oil and gas, pharmaceutical and medical devices, retail, transportation and logistics, and utility companies. As you can see from this list, IoT-based solutions can be applicable to almost every industry.

We suspect that if you work in one of these industries, you might immediately want to jump to that subsection in this chapter. However, many companies that grow adept at building IoT solutions begin to look beyond their industry for expanded business opportunities. So, you might find value in understanding what is top of mind in industries outside of where you work today.

Agribusiness Examples

Agribusiness refers to farming-related activities that include the growing and harvesting of crops, the nurturing of livestock, and the delivery of these products to market. IoT-related agribusiness applications that are deployed include

- Automated guidance of equipment used in the farm field for plowing, planting, fertilizing, irrigating, and harvesting

- Data collection from sensors in the field or drones capturing images that are analyzed to determine soil conditions (such as moisture and nutrient content), crop health, and crop maturity

- Livestock data collection that reports on their health and is used for changing feeding schedules and mixtures, for managing environmental conditions, and for suggesting optimal mating timing

- Coordination of transportation and logistics management of equipment and vehicles that transport the harvest or livestock to market

Automotive Examples

Robotics in automotive plants have relied on sensors and embraced IoT concepts for many years. These robots are involved in the manufacturing of key parts and in the assembly of vehicles.

Today, IoT is playing an increasing role in the driving and operation of vehicles in the following ways:

- Navigation of automobiles and trucks including automated parallel parking, detection of nearby obstructions that could cause damage, and self-driving vehicles with minimal driver intervention required

- Vehicle predictive maintenance and problem determination

- Scheduling of servicing based on driver usage of the vehicle

Aviation Examples

Commercial and military aircraft contain hundreds of sensors today. Until recently, while a limited amount data was transmitted to the ground while the aircraft was in-flight, the remaining massive data volumes gathered during a flight were downloaded after the aircraft reached an airport in preparation for later detailed analysis. Since more analysis is now possible onboard and transmission bandwidths and data compression techniques continue to improve, expectations are more, and real-time analysis and transmission will take place and drive

- Better and more timely predictive maintenance guidance, including scheduling of service during optimal portions of journeys

- Optimized flight operations including improvements in utilization of fuel

- More timely and better routing of aircraft in dense traffic patterns

- Better optimized baggage and cargo handling

- Timely on-ground determination of in-flight problems

- Improved capture of in-flight situations for simulation used in problem-solving, training, and certification

Communications and Media Transmission Examples

Communications, transmission of media assets, and other network providers increasingly rely on IoT gathered data for

- Improved network monitoring and problem determination

- Transmission line inspection (through image capture and analysis) for more timely repairs and safer inspections

- Improved preventive maintenance and service scheduling through predictive analysis

- Evaluation of potential new infrastructure and testing through digital simulation

Construction Examples

Companies involved in the construction of buildings, roads, and other infrastructure have deployed and/or are evaluating a variety of IoT-related solutions including

- Tracking of assets and people via location-based searches, used to direct people to equipment and tools and determine where equipment and tools are being used

- Safety problem identification (through image capture and analysis) such as workers appearing in danger zones, not wearing appropriate safety equipment, or operating/storing tools in unsafe states

- Monitoring of data from tools and other equipment to guide optimal usage and assure quality outcomes, speed work, and prevent damage to equipment

Consumer Packaged Goods Examples

Consumer packaged goods (CPG) companies manufacture, manage, and promote the items that we buy, marketing them through familiar brands and private labels. Such companies most closely monitor relationships with the channels that they sell their goods through. However, most now see a need to also directly connect with the ultimate buyers of their products, the consumers.

Examples of IoT-related initiatives include

- Supply chain optimization through better monitoring of supplies on-hand and in transit

- Better quality control and accountability through monitoring of the state and location of supplies and manufactured goods in transit

- Utilization of smart displays, sometimes linked to consumer personal mobile devices, to more quickly understand consumer buying behavior, promotional effectiveness, and impact of product placement in stores

Education and Research Examples

IoT-related initiatives touch all levels of education, from preschool to higher education. Some of these initiatives include

- Monitoring of facilities to optimize usage and control the environmental infrastructure

- Monitoring of campuses through cameras that enable image and video capture and automated analysis to help maintain security and enhance safety

- Monitoring of student presence in classrooms, libraries, and elsewhere to identify students most at risk of failing

- Analysis of data gathered from sensors and devices used in experiments and research

- Monitoring of campus or school inventories of supplies and the equipment in use, storage, and in transit

Environmental Controls Examples

Environmental controls are used to monitor and initiate changes to surroundings and typically focus on enabling delivery of desirable air quality, humidity, temperature, and water quality. These controls exist in homes and almost every industry. Some of the IoT-related use cases include

- Smarter programmable devices that can "learn" operational behaviors of operators (such as home and business thermostats that can learn desired temperature adjustments for certain days of the week and times)

- Smarter management of environmental controls for air and water quality to automatically react to a wide range of changing conditions

- Better optimization of cooling resources in manufacturing (e.g., more control over water or air cooling required resulting in less wasted resources)

- Enabling preventive maintenance on environmental controls through early detection of potential problems

Smart cities initiatives offer additional examples that might be familiar to you. Where environmental sensors have been installed during street lighting and similar upgrades, the data is sometimes used to help manage pollution challenges. For example, when levels of pollutants are approaching environmental warning levels, city governments can issue alerts and encourage carpooling and usage of public transportation. Traffic lights might also be adjusted to improve traffic flow and reduce local pollution where feasible.

Another focus of some smart cities initiatives is the optimization of environmental waste handling. Examples include the scheduling of pickup of waste materials based on fullness of recycling and nonrecyclable waste bins (monitored using embedded IoT devices and sensors) and optimal route planning for waste management vehicles.

Financial Banking and Trading Firm Examples

Banks and financial trading firms might seem to have less obvious reasons to take on IoT-related initiatives. Nevertheless, some have emerged including

- Tracking the presence and location of financial traders on trading floors

- Identifying the presence and location of handheld financial trading devices

- Tracking facility usage, especially within branch banks that are less likely to be frequently accessed by younger banking customers

Healthcare Payers and Providers Examples

Healthcare payers are responsible for managing and paying claims from services provided in healthcare providers. Healthcare providers deliver these services in hospitals, clinics, elderly care and assisted living facilities, offices of doctors, and outbound in patients' homes. Both payers and providers have an interest in delivery of quality services in the most optimal way possible. Some of the typical IoT-related initiatives include

- Improved patient monitoring in all treatment settings to better understand the impact of services provided and quality of care

- Referrals to closest facilities offering appropriate care through location-based solutions when contacted by patients

- Facilities monitoring for optimal future planning, utilization, and safety

- Monitoring of prescribed drug intake by patients utilizing smart devices and pills containing digestible sensors

- Monitoring of staff to assure quality of care and safety

High Tech and Industrial Manufacturing Examples

Many high tech and industrial manufacturing companies have deployed equipment capable of gathering data on the production floor for years and are now just figuring out how to utilize that data. There are a host of potential IoT-related solutions that manufacturers are pursuing including these:

- Gathering data on the number of goods manufactured and the environmental factors under which manufacturing occurred

- Early detection of improperly manufactured or assembled goods (through image recognition analysis)

- Refined robotic manufacturing capabilities requiring dexterity and speed

- Predictive maintenance analysis of equipment on the manufacturing floor and scheduling of servicing that will optimize production and minimize possible downtime

- More accurate and location-based assessment of inventory and the supply chain

- Better understanding of manufacturing processes associated with warranty claims and optimization of production that will minimize such claims in the future

- Manufacture of smart products that enable improved maintenance by the manufacturer and/or might enable the manufacturer to become a service provider as well

Insurance Company Examples

Insurance companies focus on selling policies at competitive rates in favorable risk profiles. So, IoT-related initiatives often focus on reducing risk and potential claims from policy holders. These initiatives include

- Analysis of vehicle driving behavior through data gathered from onboard sensors/devices

- Analysis of building usage and monitoring of security using data from sensors and image analysis

- Analysis of vehicle and building damage captured in images by cameras on mobile devices to determine the response needed and potential cost of claims

- Predictive risk modeling using data from sensors gathering weather, farming, traffic, and a host of other data related to possible claims that might occur

Law Enforcement and Emergency Services Examples

To be optimally effective and possibly save lives, law enforcement and emergency services must be properly routed to the right place at the right time with the right resources. Some of the IoT-related initiatives that can help solve this puzzle include

- Personnel, vehicle, and asset tracking enabled through the analysis of data collected by sensors and cameras

- Analysis of data collected by sensors and cameras in smart cities initiatives and linked to dispatchers of services

- Validation of identification through image recognition

Media Content and Entertainment Examples

Creators of media content and managers of entertainment venues want to quickly understand trends in popularity in order to deliver the right content at the right time to as many consumers as possible. Some examples of IoT-related initiatives include

- Analysis of crowd wait times in theme parks and entertainment venues to route individuals to lines that will improve their experience and optimize revenue through additional offerings sold

- Analysis of venue utilization for purposes of scheduling entertainment and venue redesign for better optimization

- Determination of media viewing habits through image capture and analysis of participants in studies

- Location-based recommendations provided to potential customers based on interests and/or presence

Oil and Gas Examples

Oil and gas companies are referred to as being "upstream" where exploration and extraction occur and "downstream" where production and delivery to customers occur. Companies that provide pipelines and other transport of extracted materials are referred to as "midstream." Many of the following IoT initiatives are relevant for all these types of oil and gas companies:

- Asset management including equipment, personnel, and safety considerations

- Optimal transportation and logistics management

- Preventive maintenance of vehicles, drilling, pipelines, and other equipment enabling optimal business performance and minimal environmental impact

- Sensor and image analysis at drilling sites enabling optimal discovery initiatives

Pharmaceutical and Medical Device Examples

Pharmaceutical and medical device companies engage in the research, testing, manufacturing, distribution, and promotion of drugs and devices. Historically, their primary target for these products were the caregivers. Today, many of the drugs are also directly marketed to consumers through advertising.

Some of the current IoT-related examples for this industry include

- Gathering of data from sensors and its analysis during research, experimentation, and clinical trials

- Monitoring of key metrics gathered by medical devices and fitness bands or smart watches that indicate the current state of patient health and provide warnings of potential future health problems

- Monitoring of medical devices for anomalies and possible need for replacement

- Tracking of proper intake of drugs that are monitored through equipment or digestible sensors

Retail Examples

Retailers frequently operate in an omnichannel world today going to market through physical stores, an online presence, and operations that deliver goods directly to consumers. IoT-related focus areas often include

- Utilization of smart displays to more quickly understand consumer buying behavior, promotional effectiveness, and the impact of product placement in stores. The displays also enable personal shoppers to more quickly gather items on shopping lists of consumers

- Inventory optimization through better monitoring of inventory on-hand and in transit from suppliers

- Better quality control and accountability through monitoring of the state and location of goods in transit

Transportation and Logistics Examples

Transportation and logistics management is relevant to a variety of companies and organizations involved in the shipment of goods and people. Examples include the airlines, trucking companies, railroads, and companies that manage ships at sea. Delivery companies often manage their own networks and resources but also rely on these companies. Governments also offer this service in the form of post offices delivering packages and parcels around the world.

Other companies that produce, manufacture, or sell goods also place significant focus here as it is an important cost of doing business and optimal management is key to maximizing sales and profits. In the military, effective transportation and logistics of equipment and personnel can be the difference in winning a battle.

It should come as no surprise that these are frequent IoT-related initiatives:

- Route optimization through the analysis of traffic patterns, crews, weather, and equipment, the required movement of goods and people, and the priorities under consideration (speed, cost, cost of delay, etc.)

- Service optimization through analysis of data gathered from equipment that indicates a need for preventive maintenance

- Warehouse optimization by understanding the location of inventory and supplies in storage and whether to source/deliver from or to primary or secondary warehouses or direct ship

- Network planning utilizing the results from previous optimization efforts to develop more optimal transportation paths (often evaluating multiple possible modes of transportation)

- Safety enforcement through the monitoring of vehicle operators for unusual behavior (lack of attention, lack of rest, speeding, improper lane usage) and the implementation of automated safety controls (such as Positive Train Control)

Utility Company Examples

Utility companies provide the electricity, natural gas, and water that we use to power, heat, cool, and comfortably live and work in our homes and businesses. IoT-related data initiatives in utility companies include

- Gathering and analysis of usage data from smart meters to understand resource utilization, outage situations, and predict demand

- Analysis of data gathered in plants and treatment facilities used to optimize and manage production and processes in a safe manner

- Optimal management of crews, vehicles, and other assets to maintain levels of service and maximize safety

- Utilization of image capture and analysis of images gathered by drones dispatched to remote and dangerous locations of transmission lines, pipelines, and facilities to troubleshoot existing problems and determine maintenance needs

IoT Reference Architectures

A variety of IoT reference architectures are widely promoted by standards organizations, the open-source community, and vendors that provide components and platforms. While we'll focus on the Microsoft Azure architecture in this book, gaining an understanding of other reference architectures is useful, especially when we use them to assess functional capabilities that are required in any IoT architecture.

Many of the early reference architectures emerged from efforts in the Industrial Internet of Things community. ISA-95 is an ANSI standard from the International Society of Automation that is useful in defining automated interfaces between enterprise systems and control systems. Table 1-1 illustrates the levels defined in ISA-95 including the typical systems or functions at each level.

Table 1-1. *ISA-95 enterprise and systems/function levels*

Level	Level Name	Decision Timing	Typical Systems/Functions
5	Governance and planning	Months/years	Quality management, knowledge management
4	Business systems	Days/weeks	Financials, supply chain, CRM
3	Operations management	Minutes/hours	Machine learning
2, 1, 0	Control and assets	Sub-second	Connected IoT devices

The Industrial Internet Consortium (IIC) breaks its reference architecture into slightly different functional and system areas called domains. The five domains are defined as follows:

- **Control Domain.** Functions performed by devices, sensors, and actuators at the edge, communications that occur among them, and management required

- **Operations Domain.** Functions that operate equipment in the control domain including provisioning and deployment, management, monitoring and diagnostics, prognostics (predictive analysis), and optimization

- **Information Domain.** Functions that gather data from the control domain and elsewhere into business systems (ERP, CRM, MES, etc.), custom applications, and analytics and data management systems

- **Application Domain.** Application logic or functions for performing high-level business functions

- **Business Domain.** Business processes and procedures typically found in ERP, CRM, and other systems

The way in which these domains and IoT devices operate together in an implementation is illustrated in Figure 1-2.

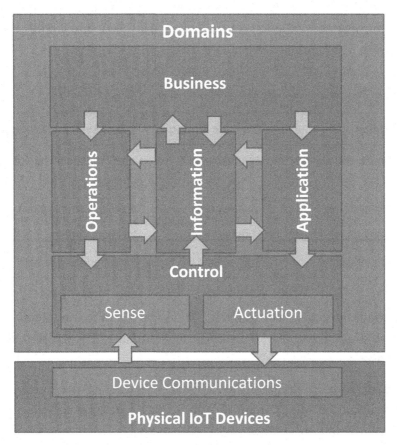

Figure 1-2. *IIC domain interrelationships and IoT devices*

There are many other reference architectures from other standards bodies and consortiums, such as the Open Software Foundation, that you might find are worth further investigation. Of course, these architectures continue to evolve as the capabilities in IoT solutions grow. But next, we start to look at how you might incorporate these concepts in your existing IT architecture.

How IoT Fits in Your IT Architecture

If you are new to IoT but have worked with IT architecture for years, you are likely familiar with traditional batch-oriented infrastructure patterns. Data in online transaction processing systems service business areas such as financial operations, supply chain and distribution, human resources, and customer relationship management. Such systems can also include unique solutions required in the industry that the company operates within. The data is structured and fits neatly into rows and columns; hence, it is stored and accessed in relational databases.

For analysis of data that crosses lines of business and requires history dating back months or years, data warehouses and/or data marts provide a place to access such data within relational databases using business intelligence tools or directly using SQL. These data warehouses and data marts are populated with data using batch extraction, transformation, and loading processes (ETL) in systems between the sources and targeted systems. They are sometimes populated using batch extraction, loading, and then transformation processes executed within the targeted data warehouses and marts (ELT). Figure 1-3 represents this architecture.

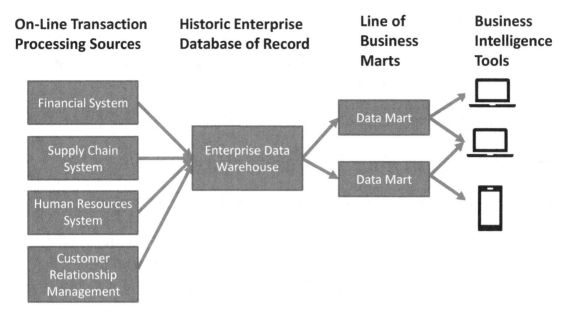

Figure 1-3. *Traditional batch-oriented data warehousing architecture*

Most of the IoT-related use cases we described in the previous section of this chapter share characteristics that drive a need for new capabilities and components beyond those that our traditional technical architecture can provide. These components must handle

- Streaming data that is generated in semi-structured format by sensors and devices at the edge of the footprint

- Incoming events that grow dramatically as the number and capabilities of the sensors and devices deployed at the edge grow – and these events must land in backend components reliably as either real-time or frequent batch input

- Storage and management of massive amounts of this streaming data enabling the analysis of patterns in the data and determination of the most appropriate machine learning models that can be deployed in the backend systems or at the edge

These requirements are very different from the requirements that drove the creation of data warehouses that are deployed using relational databases. The new architecture that emerged is often described as a Lambda architecture and consists of both real-time data feeds (a speed layer) and batch data feeds. Figure 1-4 illustrates a conceptual view of the processing and data management systems present in the architecture.

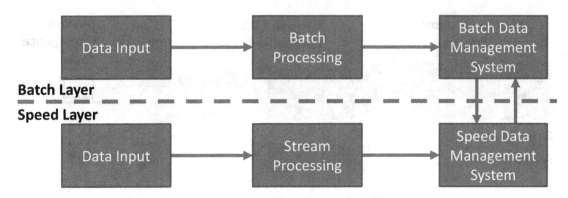

Figure 1-4. *Simplified Lambda architecture representation*

Note   Where an architecture must be defined and only streaming semi-structured data is present, just a speed layer is needed. All data is appended to a speed data management system (e.g., a NoSQL data store). This variation is referred to as a Kappa architecture.

Figure 1-5 illustrates in more detail the components that are typical in an IoT Lambda architecture. There can be many variations in the components and patterns present. The existence of legacy components, such as the presence of historians or limited networking options, can be a factor in the components that are included in this architecture. Certain functionality requirements and skills of frontline workers, developers, data scientists, and IT also influence the components selected in deployment strategies.

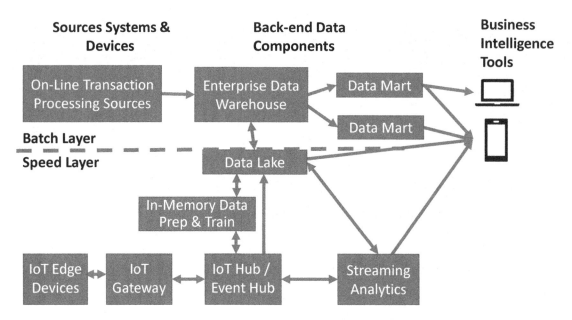

Figure 1-5. *Lambda architecture with IoT components*

IoT-related components present in the Lambda architecture diagram shown here include the following:

- **IoT Edge Device.** Remote devices that gather data and transmit it over a local area network or Wi-Fi to an IoT gateway where it is transmitted into the cloud. More sophisticated edge devices (sometimes called thick devices) can apply analytics and machine learning algorithms to incoming data.

- **IoT Gateway.** A connection point that gathers data from IoT devices and transmits it to cloud-based backend resources through public or private networks.

- **IoT Hub/Event Hub.** Both types of hubs are designed to handle a high volume of incoming messages from IoT Edge devices and support industry standard protocols such as AMQP and HTTPS. IoT Hubs additionally can provision and manage IoT Edge devices and sometimes have additional capabilities such as having support of additional transport protocols.

- **Streaming Analytics.** A real-time event processing engine used in applying machine learning algorithms and analytics to incoming streaming data feeds.

- **In-Memory Data Preparation and Training.** Spark-based solutions used to prepare data and/or perform experiments that train models in a low-latency environment.

- **Data Lake.** A location where data is stored in its natural format (usually semi-structured) in file systems or blob storage, most often leveraging Hadoop or other NoSQL data management engines.

In Figure 1-5, we show a bidirectional exchange of data between the data lake and the data warehouse. This is typical where data from one of the data management systems is needed in the other for query and reporting or machine learning training. In Chapter 2, we'll describe Microsoft's products that align to this architecture footprint.

All the new backend components are typically deployed as cloud-based services. Components you are likely to find deployed in the cloud include the IoT Hub/Event Hub, streaming analytics engine, in-memory data preparation and training solution, and the data lake. Other traditional backend components, such as the data warehouses and data marts, are sometimes relocated to the cloud, especially when replacements for a previous generation of components are sought. Dashboard and reporting business intelligence tools are also frequently cloud based.

Why Cloud Computing and IoT

When cloud computing was first introduced, the primary justification to move infrastructure to the cloud often cited was reduced cost in comparison to on-premises deployment. The cost of storage of large amounts of data is very low in most cloud-based solutions, and processing is charged for only when applications and tools utilize processing resources. However, many organizations now mention other primary motivators in moving away from considering an on-premises backend IoT infrastructure deployment.

For organizations deploying IoT in order to innovate and provide business solutions like those that we previously described, shortening the time to solution implementation can be of critical importance. On-premises deployment involves acquisition of servers and storage, software components, and networking resources. Once acquired, these resources must be installed, configured, and tested. IT must also be properly trained to manage, support, and optimize these resources. After meeting these prerequisites, development of the solutions can begin on the eventual production platforms.

In most organizations, getting the needed components in place to begin development can take 6 months or more. Utilizing cloud-based resources eliminates much of this preparation work as the new backend components can be easily spun up in minutes.

Table 1-2 denotes the resources that IT configures and manages in an on-premises deployment. This table also denotes IT responsibilities for the three types of cloud-based deployment: Infrastructure as a Service (IaaS), Platform as a Service (PaaS), and Software as a Service (SaaS). In all these scenarios except the SaaS scenario, you are responsible for managing IoT devices at the edge and their remote networks. In the SaaS scenario, managing IoT devices and remote networks is often a shared responsibility with the SaaS provider (and we denote that in the following table by an asterisk).

Table 1-2. *IT responsibilities in on-premises and cloud-based deployment*

Components Configured and Managed by IT	On-Premises Backend	Infra. as a Service (IaaS)	Platform as a Service (PaaS)	Software as a Service (SaaS)
IoT applications in data center	X	X	X	
IoT data in data center	X	X	X	
IoT data management Platforms in data center	X	X		
IoT data center middleware	X	X		
Data center operating systems	X	X		
Data center virtualization	X			
Data center servers and storage	X			
Data center networking	X			
Data center environment (power, etc.)	X			
IoT devices and remote networks	X	X	X	*

In IaaS deployment scenarios, the cloud vendor is responsible for the data center environment, networking, servers and storage, and virtualization. You remain responsible for updating and managing software and managing data in the data center above the virtualization layer. Multi-vendor software components are typically deployed, and integration among software components must be carefully considered.

In PaaS deployment, the cloud provider additionally takes on updating and management of data center operating systems, middleware, and data management platforms. Much of the focus of Microsoft's IoT reference architectures is on PaaS components, as you will see in Chapter 2.

SaaS solutions are typically sold by cloud provider partners who built their offerings upon the cloud vendor's IoT reference architectures. Examples of such offerings today come from producers of vehicles, controls, manufacturing equipment, manufactured products, and healthcare monitors. Many these products are bundled with appropriate embedded software for operating and managing the vehicle or device. Increasingly, the companies that produce these products also offer maintenance services that rely on data gathered from the equipment.

In addition to these cloud-based deployment models, there are emerging solutions that combine PaaS-style deployment with some aspects of SaaS. Microsoft refers to its offerings that cross these boundaries as solution accelerators. The accelerators spin up needed PaaS components and can provide a starting point for deploying solutions that monitor devices, create a connected factory, perform predictive maintenance on equipment, or test IoT solutions on simulated devices.

Cloud-based platforms have other benefits as well. Deploying to the cloud enables flexibility in deployment by offering a variety of reliability and availability options. Backend platforms are secured in ways not easily replicated in an on-premises deployment. And, you can rapidly scale the platforms when needed.

The cloud is also ideal for testing new IoT-based business solutions that might not prove to be justifiable. Cloud resources can easily be shut down if the project doesn't move forward without a penalty of having made a huge investment in infrastructure prior to the testing.

Other IoT Concepts and Considerations

IoT devices are of varying sophistication and capabilities. Low-power sensors might simply capture and transmit data. Powerful edge devices often contain processing units, memory, and storage and feature the ability to host an operating system and applications. The more sophisticated edge devices can process analytic and machine learning workloads that can drive immediate responses when changing conditions are detected at the source.

When specifying the remote IoT devices that you will be deploying and managing, understanding the networking options available is an important consideration. The IoT devices might communicate directly device to cloud (D2C) and receive feedback and updates from the cloud to device (C2D). More commonly, several devices in a location will transmit to an IoT gateway in a hub-and-spoke fashion as mentioned earlier in this chapter. In some scenarios where IoT devices are widely dispersed, a mesh framework is deployed as some of the devices also store and forward messages from outlying devices. A mesh containing intelligent edge devices capable of performing analytics at the device is sometimes described as a fog computing environment.

IoT devices in a hub-and-spoke or mesh deployment are most often connected using a physical connection such as Ethernet, Bluetooth, or Wi-Fi. Low-cost LP WAN technologies for devices with limited capabilities and battery life provide an alternative in some situations.

As this book was being published, there was much anticipation about the impact of 5G networking. Early Wi-Fi deployment of IoT devices utilized 3G or 4G networks. The 5G networks promise greater speed, capacity, and reliability, enabling more sophisticated exchanges between IoT devices, including intelligent edge devices, as well as communications to cloud-based services.

Communications to cloud-based services historically leveraged the Internet. However, as concern grew regarding the security of the IoT infrastructure, many organizations have chosen to connect devices to the cloud through private networks.

Messages are transmitted over these networks using message transport protocols. Some of the messaging protocols you are likely to find supported on the IoT devices that you deploy include

- AMQP (Advanced Message Queuing Protocol)

- MQTT (Message Queue Telemetry Transport)

- AMQP or MQTT over web sockets

- HTTPS (Hypertext Transfer Protocol Secure)

- Custom protocols

The devices that you choose should support a common protocol so that they can communicate with each other, and that protocol should also be supported in the cloud-based infrastructure of your cloud vendor. Though this might seem to be a trivial point, coordination between IT and the purchasers of the devices is vital to assuring success. In scenarios where this coordination is lacking, the resulting infrastructure can become needlessly more complex or early device investments could be abandoned.

Securing the infrastructure extends beyond just the network concerns. The National Institute of Standards and Technology (NIST) defines a security life cycle for an entire IoT infrastructure in its Risk Management Framework (RMF). The closed-loop process in RMF includes the following steps:

- Categorize devices and systems

- Select security controls

- Implement security controls

- Assess security controls

- Authorize devices and systems

- Monitor security controls

ISA 99 further defines relevant security assurance levels (SALs) designed to measure adherence to security goals. Target SALs are the security-level goals that are to be achieved by the IoT architecture. Design SALs are planned security levels in components and across the proposed architecture. Achieved SALs are actual measured levels of security achieved in deployment. Capability SALs are levels that can be achieved through configuration of security options in components.

Key criteria evaluated in each SAL include

- Access controls through identification and authentication

- Use control through specified privileges

- Data integrity

- Data confidentiality

- Data flow restrictions

- Time to respond to a threat event

- Resource/component availability (that could be impacted by an attack)

As we explore the various Microsoft components that might be deployed in an IoT architecture solution, we'll also take a closer look at meeting these criteria through capabilities available in Azure and at the edge.

A concept you might encounter as you devise IoT-based architectures that deliver needed business processes and solutions is the notion of a "digital twin." Organizations often build prototypes of the architecture we've described earlier in this chapter and virtually simulate the sort of data that will be gathered from sensors and devices to illustrate what solutions will deliver. Such simulations can be useful in determining which IoT devices should be deployed, or where additional sensors need to be added to devices already in place.

An Evolution in Needed Skills

By now, you likely realize that the skills that are required to successfully create and deploy an IoT-based solution, especially one that you heavily customize, are quite diverse. Possessing an understanding of the business solutions that IoT can enable and the business requirements that align to a need for IoT-based solutions in your company is required.

If you are new to IoT and to semi-structured data feeds, you will likely need to also acquire new technical skills in your company. The availability of such skills could influence the architecture that you propose. Some of the key areas and skills that will be required include

- **IoT Devices.** Understanding human to machine interfaces, device networking, device security, and device management; understanding the capabilities of such devices including programming options and deployment of analytics and machine learning at the edge

- **Streaming Data Feeds.** Understanding deployment strategies for IoT Hubs or Event Hubs, deployment of streaming analytics solutions used in the application of real-time machine learning applications, and strategies for securing data in motion

- **Semi-structured Data Management Engines.** Understanding usage and deployment of Hadoop clusters or other NoSQL databases to appropriately sized and configured systems, when to apply in-memory (Spark) processing, and data governance and security

- **Machine Learning and Artificial Intelligence.** Building data scientist skills for solving problems that require machine learning and artificial intelligence including modeling, programming, and deployment skills

- **Cloud-Based Solution Deployment and Management.** Understanding design, rollout, management, and securing of IoT backend solutions in a cloud-based environment

- **IoT Infrastructure Integration to Legacy Systems.** Understanding data integration strategies and approaches leveraging new IoT and legacy systems

Defining, designing, and implementing IoT-based solutions often follows a "design thinking" paradigm. We discuss design thinking as an approach in Chapter 8. The paradigm is a rapid cycle of research, problem definition, ideation, prototype building, and testing in an iterative fashion. Such an incremental approach is aligned with popular methodologies used in the cloud-based deployment of solutions and consistent with a modern DevOps approach. Possessing skills related to this approach and cloud-based deployment and management are also needed.

In this chapter, we outlined what these solutions might look like in a generic IoT architecture and some additional considerations. In subsequent chapters, we'll explore the key Microsoft components that can play a part in an IoT architecture and some of the possible architecture variations in more detail.

CHAPTER 2

Azure IoT Solutions Overview

Microsoft has three public cloud-based services offerings. Key components in the backend of Microsoft-based IoT solutions reside in the Azure cloud. Azure provides a platform for development and deployment of highly customized IoT solutions and for deployment of IoT applications and solutions developed by Microsoft's partners. Thus, there are examples of IoT implementations deployed in Azure using IaaS components, PaaS components, and SaaS components.

The other two Microsoft cloud-based services are Office 365 (also included in Microsoft 365) and Dynamics 365. Office 365 is a cloud-based modern workplace SaaS offering that features a variety of popular tools including Excel, PowerPoint, Word, OneNote, OneDrive, and Power BI for personal productivity, and collaborative tools such as SharePoint, Outlook, and Teams. Dynamics 365 is a suite of business applications that deliver solutions for customer sales, service, field service, finance and operations, marketing, and talent management. Microsoft's cloud-based modern workplace components and business applications often provide some of the functionality required in IoT solutions.

The Azure public cloud is available in data centers in over 50 regions around the world. For IaaS implementations, Azure provides the underlying compute, storage, and networking required. In its PaaS offerings, Azure additionally offers artificial intelligence (AI), analytics, data services, IoT components, integration components, media and content delivery network (CDN), DevOps and developer environments, compute and container services, and web and mobile development and deployment environments. Azure features an extensive management and security framework and the tools needed to support all these implementations.

© Robert Stackowiak 2019
R. Stackowiak, *Azure Internet of Things Revealed*, https://doi.org/10.1007/978-1-4842-5470-7_2

In this chapter, we introduce Microsoft components relevant in an IoT deployment that reside in Azure. We also describe Microsoft technologies deployed in devices at the IoT edge. The chapter includes the following major sections:

- Microsoft Azure and IoT PaaS

- Non-Microsoft components in Azure IoT

- IoT SaaS solutions in Azure

- Azure deployment and management

- Microsoft intelligent edge

- Choosing the right component model

We begin this chapter by focusing on Azure PaaS components deployment scenarios in IoT solutions.

Microsoft Azure PaaS and IoT

In Chapter 1, we introduced the IoT reference architecture shown again here as Figure 2-1. Within the Microsoft Azure cloud, the following speed layer components can be deployed as PaaS components: the IoT Hub/Event Hub, streaming analytics engine, in-memory data preparation and training, and the data lake.

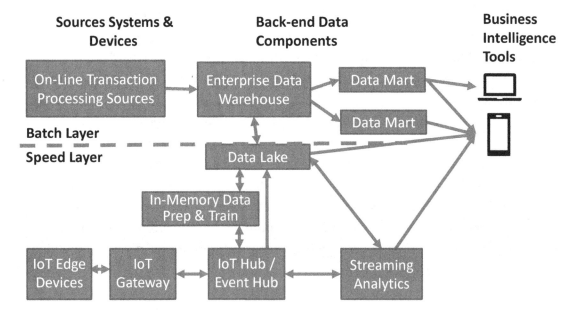

Figure 2-1. *IoT reference architecture diagram*

Microsoft Azure offerings aligned to these components include the Azure IoT Hub, Azure Stream Analytics, Azure Databricks, and data lake solutions that can include Azure Data Lake Storage (ADLS), HDInsight, and/or Cosmos DB. For analysis of time series data first landed in the Azure IoT Hub, Azure Time Series Insights is added to the architecture. Figure 2-2 illustrates where these offerings fit in the IoT architecture diagram.

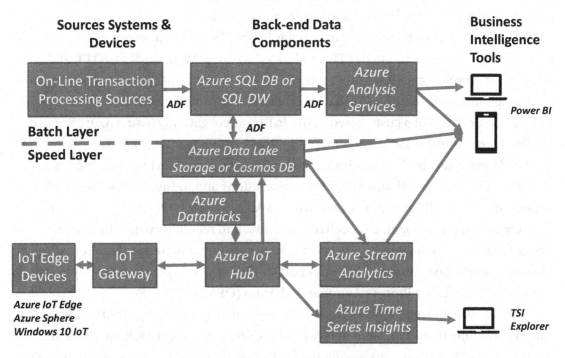

Figure 2-2. *Microsoft components in the IoT architecture*

In the batch layer, the enterprise data warehouse and data marts can also be deployed as PaaS components. Figure 2-2 also illustrates where many of these fit, including Azure SQL Database, Azure SQL Data Warehouse, Azure Analysis Services, Azure Data Factory (ADF), and Power BI.

The diagram in Figure 2-2 also notes the Microsoft offerings at the edge. These include Azure IoT Edge, Azure Sphere, and Windows 10 IoT. We will discuss the edge components later in the chapter.

Azure IoT Hub

Microsoft recommends deploying its Azure IoT Hub cloud service to enable connection of IoT edge devices to the Microsoft Azure cloud. IoT Hubs are capable of ingesting billions of events per day and support integration with Azure Stream Analytics, Azure Time Series Insights, Databricks, Azure Data Lake Storage, and HDInsight. The IoT Hub utilizes Microsoft's Event Hub technology for telemetry flow.

The IoT Hub supports a variety of popular IoT protocols for queueing and transmission of data including HTTPS, AMQP, AMQP over WebSockets, MQTT, and MQTT over WebSockets. Other protocols can be handled through protocol conversion at the edge within the Azure IoT Edge or by performing protocol conversion in the cloud through deployment of a customized Azure IoT protocol gateway (using Azure Service Fabric, Azure Cloud Services worker roles, or Windows Virtual Machines).

The Open Platform Communications (OPC) Foundation, of which Microsoft is a member, collaborates with many industry associations and industry standards bodies in defining IoT specifications. The OPC Unified Architecture (OPC UA) specifications were created to ensure open connectivity, security, and reliability where industrial devices and systems are linked. The specifications are documented in the International Electrotechnical Commission (IEC) standard IEC 62541. OPC UA was also adopted by The Open Group Open Process Automation Forum (OPAF).

OPC UA as deployed in the Microsoft IoT architecture supports publish-and-subscribe connections and client-server connections with the IoT Hub. In a typical configuration, OPC UA servers are deployed at the edge, and OPC Proxy and Publisher modules are deployed in the Microsoft IoT Edge.

The IoT Hub also provides other key functionality in the architecture. It is used for managing devices and device twins, and for identity and authentication, file upload from devices, device provisioning, and cloud-to-device messaging. Authentication is through SAS tokens, individual X.509 certificates, or an X.509 Certificate Authority. An IoT Hub can support up to 100 devices running Microsoft's IoT Edge.

Note   A device twin is a JSON document maintained in the IoT Hub that contains device-specific metadata, configurations, and conditions. It is also used when synchronizing workflows operating between the IoT Hub and edge devices (such as when firmware updates are performed).

The support of bidirectional communications enables the sending of commands, policies, and cloud-generated intelligence back to edge devices. You can store, synchronize, and query device metadata and state information, set device state, and automatically respond to device state changes using message routing integration.

Azure Digital Twins

A digital twin provides a means to represent the location of a device in the physical world. Azure Digital Twins are deployed using Azure IoT Hub technology as a foundation.

Spatial intelligence graphs are used to provide a virtual representation of the real world. Relationships between people, places, and devices can be modeled through the schema. For example, you might represent a building by defining tenants, customers, regions, building names/addresses, floors, areas within floors, and devices. You can then query data within these contexts (e.g., by location).

An example usage of a digital twin would be for processing sensor data that indicates the environmental conditions at a manufacturing site. The Azure Digital Twin would be used to validate, match, compute, and dispatch the telemetry data. Computation is executed from within user-defined functions. Using the spatial intelligence graph, you can then query data sent to the Azure Digital Twin by sensor location.

Azure Stream Analytics

Azure Stream Analytics provides an event processing engine that enables the examination and analysis of high data volumes streaming from devices. The analysis can include the extraction of information, patterns, and relationships. Actions can be triggered downstream as a result of this analysis.

Stream Analytics ingests data from the Azure IoT Hub. Stream Analytics jobs then process the data using SQL transformation queries to filter, sort, aggregate, and/or join the streaming data. The data output type is specified. Data can be sent to queues that then trigger alerts or workflows. It can be visualized in real time through tools such as Power BI. Data can also be sent to the data lake for the training of machine learning models.

Azure Time Series Insights

Time series data represents how conditions, assets, or processes change over time. Gaining an understanding of such changes to trigger actions is often the point of IoT solutions. This type of streaming data typically includes a timestamp and arrives in the order in which it was gathered.

Azure Time Series Insights parses data in JSON messages and structures that arrive from the Azure IoT Hub into clean rows and columns. It indexes the data in a columnar store and stores the data in memory or SSDs for up to 400 days (hence this is sometimes referred to as a "warm" data source given the mix of real-time and historical data). Data can be queried and visualized using the Time Series Insights (TSI) Explorer.

Azure Databricks

Azure Databricks is an Apache Spark-based analytics platform used for in-memory data preparation and in the training of machine learning models. In an IoT solution footprint, raw streaming real-time data can be ingested directly from the IoT Hub into the Databricks cluster. The data usually eventually lands in a data lake for persistent storage. Data can also be extracted from persistent storage such as Azure Data Lake Storage, Cosmos DB, the Azure SQL Data Warehouse, and non-Azure data store sources.

The collaborative workspace provided by Databricks enables the exploration of data; programming development in notebooks; data visualization through popular programming packages and toolkits such as Matplotlib, ggplot, and D3; and creation of dynamic reports. Programming languages supported in Databricks include Python, R, Scala, and SQL.

Though you can designate a fixed number of workers, autoscaling of clusters assures that a proper number of workers are always present to execute jobs. You simply specify a minimum and maximum number of workers and turn autoscaling on; clusters are appropriately sized automatically. When jobs are run, if certain parts of the pipeline are more computationally demanding, Databricks will add additional workers during these phases and remove them when no longer needed.

Azure Data Lake Storage

At the time this book was published, Microsoft had recently introduced Azure Data Lake Storage Gen2. This represented a converging of capabilities in two previously available storage services – Azure Blob Storage and Azure Data Lake Storage Gen1.

Azure Blob Storage provides general-purpose object storage and is noted for providing low-cost tiered storage. It was frequently considered adequate for smaller data lakes. Azure Data Lake Storage Gen1 added file system semantics, directory, and file level security and was usually preferred in larger implementations.

By converging these capabilities, Azure Data Lake Storage Gen2 gains Blob Storage foundation cost effectiveness to a namespace that organizes files into a hierarchy of directories containing underlying objects. POSIX permissions can be set on the directories and files. Access control lists (ACLs) and other security extensions are also supported.

Data access is more performant than in the previous generation. The Azure Blob File System (ABFS) driver is optimized for analytics. Data can be accessed in storage using the ABFS driver from Azure Databricks or HDInsight. Data in Azure Data Lake Storage Gen2 can also be accessed using versions of Apache Hadoop, Cloudera, and Hortonworks that support ABFS.

Azure HDInsight

Azure HDInsight is Microsoft's cloud-based PaaS Hadoop environment in partnership with the Hortonworks Data Platform (HDP). Today, it is most frequently deployed on Azure Data Lake Storage Gen2. Optimized clusters can be created for Apache Hadoop, Apache Spark (for in-memory caching/processing and stream processing), Apache Hive Low Latency Analytical Processing (LLAP), Apache Kafka (enabling real-time streaming messaging), Apache Storm (for distributed stream processing computation), Apache HBase (providing a distributed non-relational database deployable in Hadoop), and Machine Learning (ML) services.

Other open-source components are also present in HDInsight clusters. These include

- **Apache Ambari.** An open-source Hadoop cluster administration tool

- **Avro.** A data serialization framework often used for data exchange services in Hadoop

- **Apache Hive.** A SQL-like query interface to data stored in Hadoop

- **HCatalog.** A storage management layer in Hadoop that exposes Hive metadata to applications

- **Apache Mahout.** Open-source distribution of collaborative filtering, clustering, and classification machine learning algorithms

- **Apache Hadoop YARN.** Automates assignment of system resources for applications and schedules and monitors jobs

- **Apache Phoenix.** An open-source massively parallel relational database engine that utilizes HBase as its store

- **Apache Pig.** A platform for data analysis, designed for parallelization, that provides a programming dialect (Pig Latin)

- **Apache Sqoop.** A bulk data transfer utility used to move data from non-Hadoop data stores (e.g., relational databases, NoSQL databases) into a Hadoop Distributed File System

- **Apache Tez.** A component library that enables developers to create Hadoop applications that integrate natively with YARN

- **Apache Oozie.** A workflow scheduler for Hadoop jobs

- **Apache Zookeeper.** Provides a distributed configuration service, synchronization service, and naming registry

Default programming languages supported include Java, Python, .NET, and Go as well as several Java Virtual Machine (JVM) languages. Pig Latin for Pig jobs and HiveQL and SparkSQL are also supported. Typical development environments utilized include Visual Studio, the Visual Studio Code editor, Eclipse, and Intellij. Notebooks used in developing, debugging, and running machine learning scripts include Jupyter and Zeppelin.

Note   With Azure Data Lake Storage Gen1, HDInsight or Azure Data Lake Analytics (ADLA) could be deployed as environments. Azure Data Lake Analytics provided a U-SQL query language interface. However, ADLA was not made available for Azure Data Lake Storage Gen2.

Cosmos DB

An emerging popular alternative to deployment of Azure Data Lake Storage environments is Cosmos DB, a globally distributed NoSQL database engine. APIs available in Cosmos DB include SQL, MongoDB, Cassandra, Azure Table Storage, and Gremlin. Spark is supported for in-memory processing of data stored in Cosmos DB.

Cosmos DB can be elastically and independently scaled for throughput and storage across any number of Azure regions. Transparent multi-master replication enables 99.999 percent availability, and regional failover capabilities can also be implemented.

The datastore is schema-agnostic. Cosmos DB automatically indexes all data. Latencies are guaranteed to be 10 ms or less for reads and for indexed writes at the 99th percentile. All data is encrypted at rest and in motion, and row-level security is provided.

Other Azure Data Stores

Azure also features relational data stores and options for more traditional data warehouses and data marts that are usually fed in a batch manner. These include

- **Azure SQL Database (SQL DB).** A relational database engine that shares a common code base with SQL Server and can be deployed as part of a managed instance, a single database, or part of an elastic pool

- **Azure SQL Data Warehouse (SQL DW).** A massively parallel relational database engine for large-scale data warehousing

- **Azure Analysis Services.** Enables creation of tabular models often deployed as data mart solutions

Tools, Frameworks, and Services

Several tools often play important roles in the architecture. These include the following:

- **Azure Data Factory (ADF).** A data integration and extraction, load, and transfer (ELT) service that enables creation of data-driven workflows

- **Azure Data Catalog.** A tool used to register, tag, document, and annotate data sources through metadata

- **Power BI.** Microsoft's business intelligence tool used in the creation and analysis of reports and dashboards

Azure features a variety of options for AI development. The primary tools utilized include

- **Visual Studio.** The AI tools extension enables you to develop models deployed in Azure while providing a desktop programming interface for popular programming languages such as Python.

- **Azure ML Service.** Accessible through the Azure Portal; you have access to a modeling and deployment interface. You can also access the service through popular open-source frameworks such as PyTorch, TensorFlow, and scikit-learn. Jupyter notebooks are commonly used for programming, debugging, and running scripts.

Azure Cognitive Services are APIs, SDKs, and services that help developers build intelligent applications that can detect images and faces, detect anomalies, understand speech and language, and more. Key APIs include the following:

- **Vision.** Computer Vision, Custom Vision Service, Face API, Form Recognizer, Ink Recognizer, and Video Indexer

- **Speech.** Speech Services and Speaker Recognition API

- **Language.** Language Understanding (LUIS), QnA Maker (for easy Bot creation), Text Analytics, and Translator Text

- **Search.** Bing Web Search, Bing News Search, Bing Video Search, Bing Image Search, Bing Visual Search, Bing Custom Search, Bing Entity Search, Bing Autosuggest, Bing Local Business Search, and Bing Spell Check

- **Decision.** Anomaly Detector, Content Moderator, and Personalizer

An increasing number of these cognitive services can be deployed to intelligent edge devices in containers. As this book was published, services that could be deployed to the edge included parts of Anomaly Detector, Computer Vision, Face, Form Recognizer, LUIS, Personalizer, Speech Service API, and Text Analytics.

Non-Microsoft Components in Azure IoT

Non-Microsoft components are sometimes chosen for deployment in Azure IoT footprints. The reason for taking this approach is often because of preexisting strategies for deployment of other vendors' components. In such situations, the organization likely made an investment in software development and skills attainment tied to the component. For example, legacy ETL tools such as Informatica or Talend might already be deployed feeding on-premises or cloud-based data warehouses. The scripts that were generated might have been customized to take advantage of extended features in the data management systems that were earlier deployed.

New development using different tools and data management solutions could introduce additional costs and a learning curve. Thus, in the batch layer of the IoT architecture, we might find new development in Azure that utilizes ETL tools and data management solutions from other vendors. For example, we might find relational databases serving as data warehouses that include IBM DB2, MariaDB, MySQL, Oracle, PostgreSQL, or Snowflake.

In the speed layer, Hadoop engines from Cloudera/Hortonworks or MapR might be deployed for similar reasons. NoSQL databases such as Cassandra or MongoDB could also be present.

Note   Azure Stack is Microsoft's on-premises cloud offering that provides an Azure IaaS environment on specific server and storage configurations built by Microsoft partners such as Dell, HP, Lenovo, and others. You are more likely to find non-Microsoft software components to be part of the IoT footprint here. Deployment of on-premises cloud configurations providing the IoT backend are most often considered when limited networking availability makes connections to an off-site cloud nonviable.

Microsoft also has several IoT platform partners that utilize the IoT Hub to connect their offerings to Microsoft's Azure IoT footprint. Partners include C3 IoT, OSISoft PI, and PTC ThingWorx. Their IoT solutions sometimes leverage Microsoft data management offerings in Azure such as SQL DB or Azure Postgres. The deployment architectures from these partners typically contain components that overlap in capabilities with Microsoft Azure IoT components providing functionality in areas such as stream analytics, machine learning, and edge services.

IoT SaaS Solutions in Azure

Repeatable IoT solution architectures built upon a common set of Microsoft PaaS components are becoming increasingly common. To speed deployment of such solutions, Microsoft has created IoT Central and solution accelerators. These accelerators can also be used to provide a starting point for understanding components needed in an IoT solution since. Each solution accelerator deployment configures and spins up the necessary cloud-based services required in implementations of remote monitoring, preventive maintenance, and for other IoT solution use cases. The application code is open-sourced through GitHub.

At the time this book was written, the following solution accelerators were available at `https://www.azureiotsolutions.com/Accelerators` (and were in the process of being updated to a microservices architecture):

- **Remote Monitoring.** Collects telemetry from remote devices, monitors device condition (presented through a dashboard), and provides firmware and software update provisioning

- **Connected Factory.** Collects telemetry from industrial assets (such as PLCs, industrial barcode readers and scanners, smart meters) based on the OPC UA standard, monitors them and presents metrics in a dashboard, and enables management of the devices

- **Predictive Maintenance.** Predicts when a remote device is about to fail by applying machine learning algorithms to telemetry from those devices and provides a dashboard interface to view device status

- **Device Simulation.** Provides a means to run simulated devices that produce realistic telemetry for testing of the solution accelerators or custom IoT solutions

Some Microsoft partners have leveraged IoT PaaS components to build out their own solutions that are marketed as complete architectures, product suites, and services. Examples include Honeywell's MAXPRO Cloud providing services for connected buildings, Rockwell Automation's FactoryTalk for monitoring industrial equipment, Schneider Electric's Ecostruxure used to optimize energy and water resource utilization, and Siemens' MindSphere typically deployed in optimizing operations of industrial equipment. Other example solutions where Microsoft Azure is under the covers include connected car offerings offered by some automakers and patient monitoring and diagnosis devices offered by healthcare device manufacturers.

As noted in the introduction to this chapter, IoT solution footprints are sometimes linked to other SaaS-based cloud solutions to provide additional functionality. Microsoft PowerApps can be used to create business logic and workflow needed in custom integration between SaaS applications and IoT backend data sources.

For example, an IoT alert indicating the likely future failure of equipment might trigger a work order in Microsoft Dynamics 365. Using Connected Field Service, the right technician with the right skills can be scheduled and dispatched. They can view information on the anticipated problem and indicate when the problem has been mitigated in Dynamics.

Azure Management and Deployment

We realize that some of you might be new to Azure. In this section, we take a step back to provide a quick introduction to Azure management and deployment considerations. This is a broad topic, and entire books have been written on the subject. Here, we simply hope to highlight some areas that can help you plan your IoT deployment strategy and governance of your Azure environment.

Microsoft describes Azure as a platform built upon trust. Foundation principles for the platform include scalability and performance, manageability, resilience, availability, and security. We'll touch on these topics in the following subsections.

How you govern your environment and the technology you choose to use is determined by your business strategy and your risk profile. Technology deployment success is dependent on establishing and managing configurations, establishing policies and then monitoring and enforcing compliance, managing costs and resources, and managing security (including identities). In Microsoft documentation, outlining a governance plan for your Azure environment is described as establishing a scaffold.

Subscriptions and Resource Groups

When you utilize Azure, the resources that you consume are allocated to the subscription that you are using. Subscriptions are typically assigned to individual projects, phases in development, or by applications. Multiple subscriptions can be assigned to accounts.

Multiple accounts can be assigned to departments (typically defined by organization or geographic location) that make up your enterprise. These entities should be identified consistent with the naming standards used within your organization. They are managed using the Azure Portal.

Figure 2-3 provides an example and illustrates where subscriptions, accounts, and departments fit in this basic enterprise hierarchy.

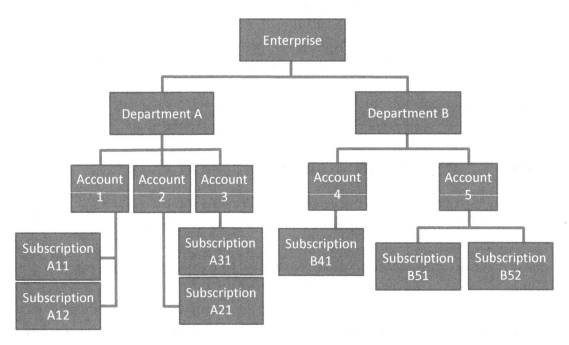

Figure 2-3. *Azure subscription hierarchy*

Hierarchies can be extended beyond those defined for billing purposes through Azure management groups. Related subscriptions can be grouped together regardless of where they are in the billing hierarchy. Common roles, initiatives, and policies can be defined across subscriptions. In addition, accounts and departments can be nested up to six levels.

The Azure Resource Manager enables the placement of common resources into groups for ease of management and billing. These resource groups typically hold the resources required by applications or other solutions that you deploy. The Azure Resource Manager can be used to enforce policies such as maintaining data sovereignty and privacy or to enable more accurate and explainable billing.

Note   For environment setup, you might also use the Azure Blueprints service. It provides a means of packaging artifacts that include resource groups, Resource Manager templates, policies, and role assignments.

Authorization in Azure Resource Manager is enabled through Role-Based Access Control (RBAC). Though there are over 70 built-in roles that are pre-defined, 4 of them provide important fundamental levels of access:

- **Owners.** Possess full access to all resources and can delegate access.

- **Contributors.** Create and manage Azure resources but cannot delegate access to others.

- **Readers.** View Azure resources.

- **User Access Administrators.** Manage user access to Azure resources.

Azure Portal

Azure applications and resource management, deployment, and monitoring are most typically performed through the Azure Portal, a web-based interface. Many management activities can also be executed through the command line interface (CLI) or through Azure PowerShell.

Figure 2-4 illustrates a typical Azure Portal dashboard view. You can view favorite available services on the left side of the dashboard (or choose to view all services). You can also use the search at the top of the Portal view to easily find services that you might want to deploy. Within the main Portal viewing area, you have access to all resources already deployed, tutorials, and workspaces. You can easily access information about the Service Health and will find the Marketplace of additional available resources.

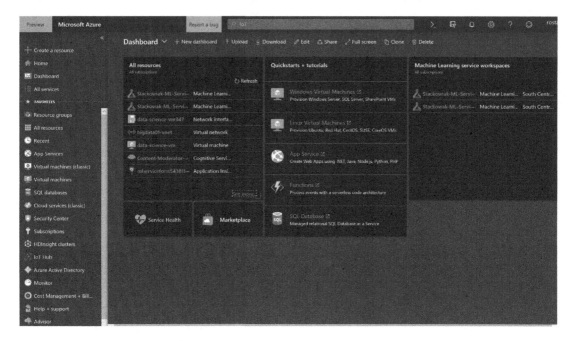

Figure 2-4. *Azure Portal*

Dashboard layouts can be customized and published. After publishing, you can share your customized dashboards with others.

Azure Monitor provides tools that collect and analyze performance and availability data for your deployed solutions. Accessible through the Azure Portal, you can use the Azure Monitor interface to set up alerts when specific conditions occur and trigger actions, query and analyze logs, or simply monitor and visualize metrics associated with your cloud resources. Metrics that can be tracked include blocked calls, client errors, data in, data out, latency, server errors, successful calls, total calls, and total errors. Data can be viewed in time segments ranging from the last 24 hours to the last 30 days.

Figure 2-5 illustrates the gathering of average latency and total calls data that was gathered over a time period of 30 days.

Figure 2-5. *Azure Monitor metrics*

Azure Advisor provides proactive and actionable best practices recommendations that guide you in improving the performance, availability, security, and cost-effectiveness of your Azure resources. Accessible via the Azure Portal, Figure 2-6 illustrates a typical view in Azure Advisor. Recommendations are noted as having high, medium, or low impact. You can then explore the recommendations provided and decide whether to implement them in each category.

Figure 2-6. *Azure Advisor calls attention to best practices recommendations*

The Azure Portal also plays a key role in managing the costs of your Azure deployment. During your initial configuration of resources needed for deployment, you will see cost choices clearly spelled out. For example, when deploying data management components, you'll have a choice of different CPU and memory classes of performance and different storage levels (premium/SSD, hot, cool, and archive).

Note   In addition to costs associated with the operation of Azure resources, you will also accrue costs when data flows out of Azure regions and between different availability zones, peered VNets, and globally peered VNets.

Through the Azure Portal, you also have access to Azure Cost Management used in monitoring and controlling Azure spending and for optimizing resource utilization based on recommendations received. Figure 2-7 illustrates costs accruing during a month up to the current date and breaks down current costs by service names, locations, and resource groups. You can additionally provide budget information and receive alerts when budget restrictions are reached.

Figure 2-7. *Azure Cost Management analysis of current month costs*

Designing for Resiliency and Availability

Resiliency is the ability of a system to recover from failures and continue to function. Availability is the proportion of time that the system is operating normally. Designing to achieve both resiliency and availability is key to meeting service level agreements (SLAs) for the backend of your IoT solution.

Microsoft's Azure architecture delivers an SLA above 99.9 percent for single virtual machines by default. The platform can take proactive automated action when potential hardware failure is detected, communicates via a Microsoft private network between regions, triple mirrors data, and has other availability design characteristics under the covers.

Resilience services available in Azure include

- **Azure Backup.** A general-purpose backup solution for workflows on virtual machines or servers

- **Azure Site Recovery.** Replication of virtual machines from on Azure region to another

47

- **Availability Sets.** Virtual machines distributed across multiple isolated cluster nodes providing protection from hardware failures within a data center

- **Availability Zones.** Distribution of virtual machines across multiple physical locations within a region where each location has independent network, cooling, and power

- **Azure Load Balancer.** Distributes traffic according to rules and health

- **Azure Traffic Manager.** Optimal distribution of traffic to services across global regions

- **Geo-replication for Azure SQL Database.** Fast disaster recovery of individual databases during regional or widespread outages

- **Locally Redundant Storage (LRS).** Replication of data to a storage scale unit.

- **Zone Redundant Storage (ZRS).** Synchronous replication of data across three storage clusters in a single region

- **Geo-redundant Storage (GRS).** Replication of data to a secondary region hundreds of miles away from the primary

Responsibilities for resiliency vary depending on the type of Azure deployment as illustrated in Table 2-1. You are responsible where an "X" is indicated in the table. The asterisk indicates a shared responsibility.

Table 2-1. *Comparison of on-premises vs. Azure resiliency responsibilities*

Components Configured and Managed by IT	On-Premises Backend	Infra. as a Service (IaaS)	Platform as a Service (PaaS)	Software as a Service (SaaS)
Database/data HA and DR	X	X	X	X
Workload/ application HA, DR, backup	X	X	X	*
Virtual machine/ OS HA, DR, backup	X	X	X	
Storage HA, DR, backup	X	X	*	
Networking HA and DR	X	*		
Power/facility HA and DR	X			
Data center environment (power, etc.)	X			

Database and data resiliency can be assured through Azure Backup and services provided by Azure PaaS databases. Workload application resiliency can be satisfied using Azure Backup and Azure Site Recovery.

Resiliency of virtual machines and operating systems can be assured through Availability Sets, Azure Backups, and Azure Site Recovery. There is 99.99 percent SLA when two or more VMs are running in separate Availability Zones within a region protecting against data center failures in comparison to a 99.9 percent SLA when just single VMs are deployed.

Storage resiliency can be satisfied through deployment of managed disk in combination with redundant storage. You might choose to configure storage as locally redundant, zone redundant, or geo-redundant depending on the level of resiliency required.

Networking resiliency is achieved through deployment of region pairs that leverage Load Balancer and Availability Zones. Region pairs provide protection for data and applications even in the event of loss of an entire region via geo-redundant storage (GRS) and Azure Site Recovery. Region Pairs and Availability Zones are also key building blocks in providing power and facility resiliency.

Azure Security Considerations

Azure security considerations include identity and access management, data protection, network security, threat protection, and security management. Key technologies present in Azure to create and manage a secure environment aligned to these considerations include

- **Identity and Access Management.** Azure Active Directory, Multifactor Authentication, Role-Based Access Control, and Azure Active Directory Identity Protection

- **Data Protection.** Encryption (disks, storage, SQL), Azure Key Vault, and Confidential Computing

- **Network Security.** VNet, VPN, NSG; Application Gateway (WAF), Azure Firewall; and DDoS Protection Standard, ExpressRoute

- **Threat Protection.** Microsoft Antimalware for Azure and Azure Security Center

- **Security Management.** Azure Log Analytics and Azure Security Center

Azure Security Center is accessible through the Azure Portal and provides a unified security management system for your Azure resources as well as for hybrid workloads. Events collected from agents and Azure are correlated in a security analytics engine, assessing whether your resources are secure. Threat prevention recommendations and threat detection alerts are raised. When those occur, you can take the recommended actions and properly provision the identified resources.

The Azure Security Center dashboard is illustrated in Figure 2-8. You can view scoring of the level of policy and compliance security, summaries of resource security hygiene and recommendations, and security alerts by severity through this view and then proceed through recommended actions.

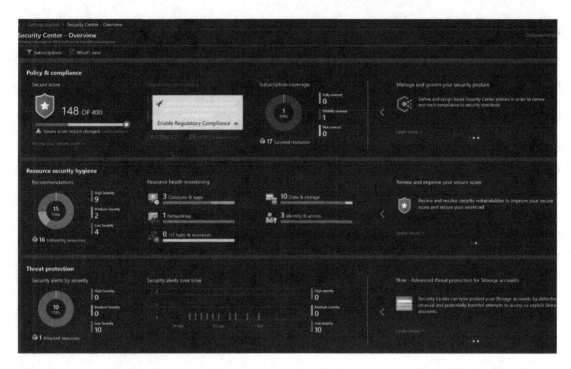

Figure 2-8. *Azure Security Center*

When planning security for you Azure-based solutions and designing, deploying, configuring, and managing them, you can get guidance from the Microsoft Trust Center (`https://www.microsoft.com/trustcenter`). There, you will find how Azure can help you meet compliance standards driven by industry and geographic requirements. You can explore the compliance manager, audit reports that are produced, and other data protection available resources such as whitepapers and documentation.

Microsoft Intelligent Edge

IoT devices gather data through sensors and transmit the data from remote locations to the Azure cloud through networks. These edge devices continue to grow in sophistication and capabilities. Today, many can run analytics and custom business logic at the edge, sometimes even when they are disconnected from the cloud. This edge device software is managed through Azure IoT Edge.

Today's sophisticated IoT devices feature CPUs, storage, and memory of varying processing power and capacities enabling deployment of operating systems. Microsoft's device operating system offerings include Azure Sphere (with Linux or real-time operating systems) and Windows 10 IoT. We introduce all of these in this section of the chapter.

Azure IoT Edge

Microsoft's Azure IoT Edge is comprised of three components: IoT edge runtime environments that run on each device, edge modules that run analytics and your custom logic, and edge cloud interfaces.

The IoT runtime environment runs on devices that support Linux or Windows. It enables software installation and updates on the device, enables secure operations and ensures that the device is operational, reports the health of modules to the cloud, and manages communications to downstream devices, between modules, and to the cloud.

Edge modules are deployed in containers and can include Azure services, third-party services, and custom code. The following Azure services can be deployed to edge devices:

- Azure Machine Learning

- Azure Cognitive Services

- Azure Event Grid

- Azure Functions

- Azure Stream Analytics

- Azure SQL Server

In addition, Microsoft announced a small footprint edge optimized data engine for the IoT Edge in 2019 named Azure SQL Database Edge. It is deployed in a container running on ARM- or x64-based devices that can be connected or disconnected from the Azure IoT backend. You can use this engine to stream, store, and analyze time series data on the device.

The IoT Edge cloud interface enables the creation and configuration of workloads in the cloud that will be run on specific devices. It is also used to provision workloads to the edge devices and monitor the workloads running on the edge devices.

Azure Sphere

Azure Sphere is a secured application environment that can be deployed in edge devices featuring a class of crossover microcontroller units (MCUs) available from Microsoft partners. A custom Microsoft Linux kernel provides a secured operating system for devices and the subset of POSIX functionality needed by some applications. Applications can be run in sandboxed containers on the device.

The Azure Sphere Security Service brokers trust for device-to-device and device-to-cloud communications. It detects emerging threats and can renew device security. Additionally, the Sphere Security Service can automate download and installation of operating system updates and ensure that the device boots only with approved software.

An alternative real-time operating system (RTOS) for these devices was announced when Microsoft acquired Express Logic, the developer of ThreadX RTOS, in 2019. ThreadX had already been deployed on over 6 billion devices including many that are highly constrained (as it requires just 2 KB in instruction area and 1 KB in RAM). The RTOS provides advanced scheduling, secure communications, synchronization, a timer, memory management, and interrupt management facilities. It supports MQTT and can connect directly to the IoT Hub.

Windows 10 IoT

Windows 10 IoT is a family of products based on the popular Windows 10 operating system for PCs and servers. Members of this IoT family include the following:

- **Windows 10 IoT Core.** A limited version of Windows 10 for less powerful IoT devices running x86, x64, ARM, or i.MX processors; enables the running of only a single application.

- **Windows 10 IoT Enterprise.** A full version of Windows 10 with additional features enabling the lockdown of IoT devices; available for devices running x86 or x64 processors.

- **Azure IoT Edge for Windows.** A runtime environment that enables deployment of Windows containers on devices running Windows 10 or Windows 10 IoT Core; used to deploy Azure services and custom logic.

- **Azure IoT Device Agent for Windows.** Enables configuration, monitoring, and management of remote IoT devices running Windows 10 from the Azure dashboard.

- **Robot Operating System for Windows.** A version of Windows 10 intended to make development of robotic applications easier; includes intelligent edge capabilities and support for Cognitive Services and hardware-accelerated Windows Machine Learning.

Choosing the Right Component Model

As IoT began to mature, footprints grew in breadth and depth. Early deployments of IoT solution backends relied on IaaS components with custom integration required between them. Diverse management tools were required to manage the entire environment, and support models were highly complex with many vendors involved.

Today, as we've illustrated in this chapter, Microsoft has an extensive array of PaaS components in Azure that are more tightly integrated. The PaaS components are all managed through the Azure Portal. Microsoft also provides software that enables critical capabilities required in devices at the edge.

This extensive footprint has enabled the introduction of Microsoft IoT solution sets that provide a starting point for deploying complete solutions. Solution accelerators found in the Azure IoT Central are increasingly featuring characteristics common in SaaS solutions. You will also find that there are many third-party solutions that rely on underlying Microsoft IoT components and are sold as packages with devices.

We'll continue to see a growing array of more SaaS-like IoT solutions in the future. How you will choose to deploy your IoT footprints will likely be driven by the devices that you purchase to meet your business and technical needs and the support and service offerings of the device or solution vendors.

Given the current diversity of devices and building block approach that is often taken when defining IoT solutions today, you likely need to gain a deeper understanding of the components required beyond the introduction that we provided in this chapter. So, in the next few chapters, we take a further look at many of the Microsoft IoT components deployed in Azure cloud-based backends and at the intelligent edge.

CHAPTER 3

IoT Edge Devices and Microsoft

The "Things" in the Internet of Things are edge devices connected to centralized computing resources through external networks. Sometimes, the devices are also connected to each other at the edge via local networks. Data might be gathered from sensors in state-of-the-art complex devices such as smart meters, industrial barcode and RFID readers, programmable logic controllers (PLCs), and robotic machinery. Sometimes, the equipment that you need to gather data from lacks sensors in critical locations. Simple sensor kits might be applied to legacy equipment lacking needed sensors. Such kits might also be installed in new locations such as upon city infrastructure, buildings, mobile vehicles, or even drones.

A variety of edge devices can be connected to the Microsoft cloud backend, most often through Azure IoT Hub (explored in greater depth in the next chapter). Communications protocols supported by the IoT Hub include the Advanced Message Queuing Protocol (AMQP), the Message Queue Telemetry Transport (MQTT) protocol, and HTTPS. However, devices using other protocols can be deployed by providing protocol translation at the edge.

We begin the chapter by describing criteria often used in the selection of sensors and devices at the edge. We then describe Microsoft's IoT Edge runtime software for devices, including using the edge device as a gateway, deploying containers to the edge, and the role of this software in securing the device. We complete the chapter with a description of the Azure IoT device catalog, noting the registration process and the available test suite used in certification of the devices by Microsoft's partners. Thus, the chapter is divided into the following major sections:

- Edge sensor and device selection
- The Azure IoT edge runtime
- The Azure IoT device catalog

55

© Robert Stackowiak 2019
R. Stackowiak, *Azure Internet of Things Revealed*, https://doi.org/10.1007/978-1-4842-5470-7_3

Edge Sensor and Device Selection

Deploying a successful IoT solution often begins with an assessment of the data needed and types of actions that must be taken in response. You might begin by evaluating whether existing edge devices gather the right data and whether sensors are in the right locations to get the measurements needed.

If the data gathered is inadequate, you could then be faced with deciding whether to retrofit existing equipment with additional sensors or perform wholesale replacement of equipment with newer IoT-ready versions. When evaluating the acquisition of new equipment, criteria considered often includes the accuracy of measurements provided as well as cost.

Additional physical considerations can include component durability, physical size, and mounting options. Critical components in the edge devices might need to function in difficult environments so understanding normal operating temperature ranges, acceptable moisture levels, the presence of acids or chemicals, and amount of electrical noise can play a part in selection.

Devices and sensors could require minimal voltage and be battery powered. Or they might need local power drops to be adequately powered. In an environment where high availability is a must, an uninterrupted power supply (UPS) could be required.

Physical and software security provided by the proposed devices and deployment in secured areas also warrants consideration. Communication protocols supported by the devices and their connectivity (wired or wireless) should be consistent with the planned overall architecture. Evaluations that touch on all these considerations are key to simplifying integration and ongoing operational management.

The support model for devices should be developed and tested as part of scalability testing prior to full deployment. Key areas that receive testing should include initial provisioning, centralized management of connections, and troubleshooting based on diagnostics. The plan should include details regarding ongoing engineering support after deployment.

Edge devices are sometimes connected directly to the Azure IoT Hub in the cloud for communications. More often, they are networked together in their remote location via local area networks (such as over Ethernet) or via wireless connections (such as Wi-Fi, 3G, 4G LTE, 5G, and Bluetooth). Remote IoT networks at the edge are connected locally to an IoT gateway device to transmit and receive data to/from the Azure IoT Hub in the cloud.

As noted previously in this book, the Azure IoT Hub supports MQTT, MQTT over WebSockets, AMQP, AMQP over WebSockets, and HTTPS for connectivity into the Azure cloud. Other protocols, such as OPC UA, are supported through protocol translation. Most organizations seek to standardize on a single protocol for device connection to the cloud to simplify their architecture.

Figure 3-1 illustrates where these protocols align to the OSI and TCP/IP communications models.

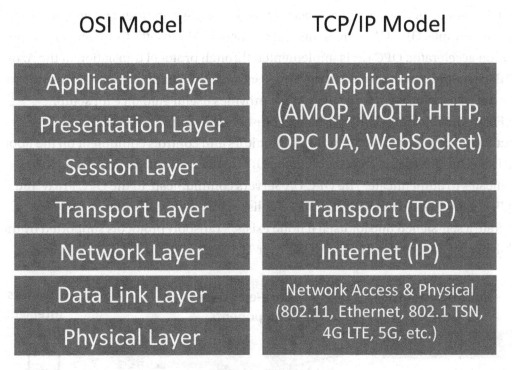

Figure 3-1. *IoT protocols in OSI and TCP/IP models*

AMQP is a popular choice where field and cloud gateways are deployed since connection multiplexing is supported. Multiple devices with unique credentials communicating using MQTT and HTTPS cannot share the same TLS connection. However, the MQTT and HTTPS protocols can run in devices with fewer resources, so they are sometimes your only option for such devices.

WebSockets come into consideration when ports needed by AMQP (using port 5671) or MQTT (using port 8883) are closed to non-HTTPS protocols. Hence you would deploy AMQP over WebSockets or MQTT over WebSockets in such a scenario if you did not want to rely on HTTPS as the protocol.

AMQP and MQTT both support a server push of messages from the cloud to the device. HTTPS devices must poll the server to determine if messages are present, thus generating additional traffic over the network. For HTTPS devices, it is recommended that such polling should be limited to frequencies of every 25 minutes or more.

AMQP and MQTT are binary protocols. Payloads are more compact when using these protocols compared to HTTPS giving yet another reason why they are often favored in IoT deployment.

For Industrial Internet of Things (IIoT) implementations, the OPC UA protocol has gained a great deal of acceptance. Microsoft was an early supporter and adopter of OPC UA. For example, OPC UA is the industrial protocol in Microsoft's Connected Factory solution accelerator. OPC UA is implemented through protocol translation in the Azure IoT Edge module, as described in the next section of this chapter. Microsoft provides a protocol translation sample in GitHub that provides useful guidance for setup.

Figure 3-2 illustrates OPC UA Servers present on multiple assembly lines. A manufacturing execution system (MES) monitors and controls equipment on the floor and is an OPC UA Client. It connects to the OPC UA Servers on the floor using X.509 certificates to authenticate. The OPC UA Servers communicate to the OPC Proxy and Publisher modules in the IoT Edge. The Publisher checks for a certificate and can generate a self-signed one for itself if none exists. Transport protocols supported by the Publisher are AMQP over TCP or MQTT over TCP.

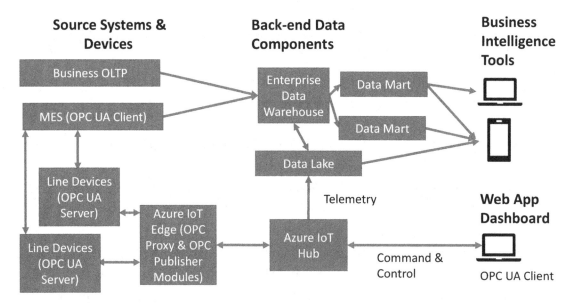

Figure 3-2. *OPC UA deployed in manufacturing*

The OPC UA Publisher also supports direct calls in the IoT Edge that provide general information, diagnostic information on OPC sessions, subscriptions, monitored items, and diagnostic information on IoT Hub messages and events. The last 100 lines of the log that is maintained can be read.

Many manufacturers of industrial equipment have adopted this protocol. In Microsoft Technology Centers in many major cities around the world, there are IIoT walls displaying some of this equipment. The walls include

- **HPE IoT System EL20.** A rugged performance-optimized edge gateway with compute capabilities designed for collecting and transmitting data in high-volume deployments.

- **Siemens SIMATIC S7-1500 Advanced Controller.** A programmable logic controller (PLC) with integrated motion controls.

- **Mitsubishi Electric MELSEC iQ-R series PLC.** A multidiscipline PLC with motion CPUs to control positioning, speed, torque, advanced synch, and other functions.

- **Leuze Electronic IoT Ready Barcode Reader.** An industrial quality barcode scanner.

- **Beckhoff IoT Controller.** Features temperature and fan control and speed sensors to demonstrate the ability to control cooling.

- **HARTING Ha-VIS UHF RFID reader RF-R310.** An industrial RFID reader.

- **Rockwell Automation Control Logic Controller.** A PLC for multi-axes motion control often used in mixing ingredients and filling containers.

- **Schneider Electric IoT Controller.** A PLC sometimes used in measuring liquid levels in tanks and for similar applications.

- **Honeywell Elster Alpha Smart AS300P.** A device that enables an advanced metering infrastructure (AMI) and manages data from electricity, gas, and water meters.

Figure 3-3 illustrates this equipment as it appears on the wall.

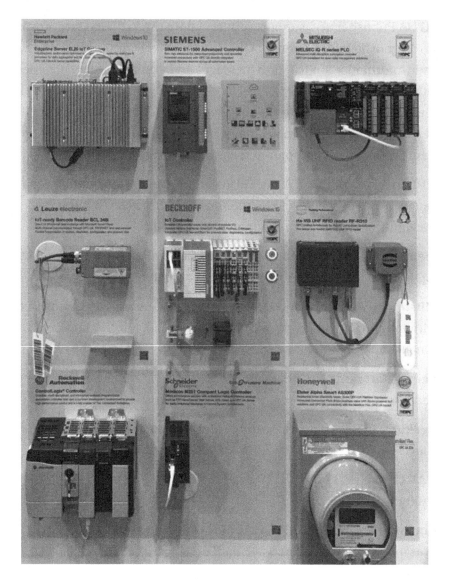

Figure 3-3. *OPC UA connected equipment in Microsoft Technology Centers*

There are other protocols that might be supported in your industrial equipment, especially in equipment that predates OPC UA. For example, Modicon (now Schneider Electric) published Modbus as a protocol in the late 1970s, and it gained some widespread adoption. Prior to OPC UA, Microsoft promoted the usage of its OPC Classic.

The Azure IoT Edge Runtime

Microsoft's Azure IoT Edge is runtime software that can perform communications and module management functions on edge devices. From an edge device, communications can occur to further downstream devices, modules present in the device, and to the Azure cloud. Management functions available include the ability to install and update workloads on the device, maintain IoT Edge security standards on the device, ensure that IoT Edge modules are running, and report on module health in the device enabling monitoring.

The IoT Edge runtime consists of two modules: the IoT Edge Hub and the IoT Edge Agent. The IoT Edge Hub serves as a local proxy for the Azure IoT Hub and supports MQTT and AMQP as protocols to the IoT Hub. It optimizes connections to the cloud and relies on the Azure IoT Hub for authentication when connections are first established.

The IoT Edge Agent instantiates modules, ensures that they are running, and reports on status back to the Azure IoT Hub. The IoT Edge Agent uses its module twin to store configuration data.

Deployment of the IoT Edge runtime begins with deployment of an Azure IoT Hub (described in the next chapter). You then register an IoT Edge device to the Azure IoT Hub and install and start the IoT Edge runtime on the device. It is then possible to deploy a module remotely to the IoT Edge device.

You can use a continuous integration and continuous delivery (CI/CD) process familiar in modern DevOps scenarios when deploying your IoT Edge. Key steps, as documented by Microsoft in GitHub, are

- Create the Azure resources.

- Set up Azure DevOps services.

- Define continuous integration (using tokens).

- Create a release pipeline and smoke test.

- Add a scalable integration test to the release pipeline.

- Monitor the devices with Application Insights.

The IoT Edge Device As a Gateway Device

IoT Edge devices can serve as transparent, protocol translation, or identity translation gateways to the Azure IoT Hub. The gateways can be used to provide downstream device isolation, connection multiplexing, traffic smoothing, and limited offline support (temporarily storing messages and twin updates at times when they cannot be delivered to the IoT Hub).

A transparent gateway simply passes communications from devices to the Azure IoT Hub for devices that support the MQTT, AMQP, or HTTP protocols. Logical device connections are multiplexed over a single physical connection using AMQP or MQTT as a protocol between the IoT Edge Runtime and the Azure IoT Hub. All connected device identities are stored in the IoT Hub identity registry, each device has its own device twin, and each device can be addressed from the cloud individually. Figure 3-4 illustrates the major components present when an IoT Edge device serves as a transparent gateway.

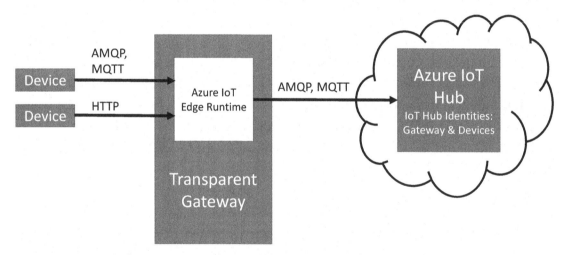

Figure 3-4. *IoT Edge device as a transparent gateway*

A protocol translation gateway can communicate with devices that support other protocols such as OPC UA, Modbus, Bluetooth Low Energy (BLE), Building Automation and Control Networks (BACnet), or other proprietary protocols. There is a single physical connection using AMQP or MQTT as a protocol between the IoT Edge Runtime and the Azure IoT Hub for the gateway only. Only the identity of the gateway device is stored in the IoT Hub identity registry, and only the gateway has a device module twin. To analyze data on a per-device basis, messages from devices must contain additional identifying information. The cloud can address only the gateway device directly, not the downstream devices. Figure 3-5 illustrates the major components present when an IoT Edge device serves as a protocol translation gateway.

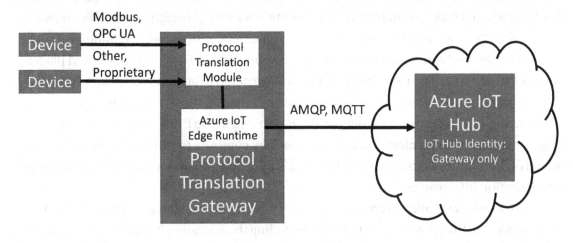

Figure 3-5. *IoT Edge device as a protocol translation gateway*

Identity translation gateways provide protocol translation but also can identify downstream devices and translate those identities to IoT Hub primitives. Logical device connections are multiplexed over a single physical connection using AMQP or MQTT as a protocol between the IoT Edge Runtime and the Azure IoT Hub. Thus, device identities are stored in the IoT Hub identity registry, each device has its own device twin, and each device can be addressed from the cloud individually. Figure 3-6 illustrates the major components present when an IoT Edge device serves as an identity translation gateway.

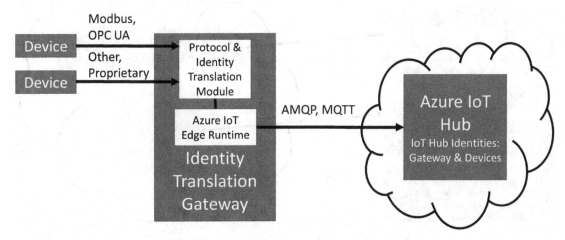

Figure 3-6. *IoT Edge device as an identity translation gateway*

Deployment of Containers

As noted in Chapter 2, Azure Machine Learning, certain Azure Cognitive Services, Azure Event Grid, Azure Functions, Azure Stream Analytics, and SQL Server can be deployed in the IoT Edge within Docker containers. Docker containers are lightweight executable software packages that include needed system libraries and settings, system tools, runtime, and code needed to run the Azure service. Code is typically built in Azure or Visual Studio and placed into a Docker image. The Docker containers are registered to an Azure container registry.

Deployment of the containers onto the device relies on the two modules present in the Azure IoT Edge runtime – the IoT Edge Hub and the IoT Edge Agent. The IoT Edge Agent instantiates modules, ensures that they continue to run, and reports the status of modules back to the IoT Hub. The IoT Edge Agent uses its module twin to store configuration information.

A deployment manifest is created as a JSON document and stored on the IoT Hub. It describes the IoT Edge Agent module twin including the container image for each module, credentials needed to access private container registries, and instructions on how modules should be created and managed. The deployment manifest also describes the IoT Edge Hub module twin defining how messages flow between modules and to the IoT Hub.

Upon device startup, an IoT Edge security daemon starts the IoT Edge Agent. This agent retrieves its module twin from the IoT Hub and the contents in the deployment manifest. The named modules are then started on the device as module instances.

Figure 3-7 illustrates these key components in deployment of containers at the edge.

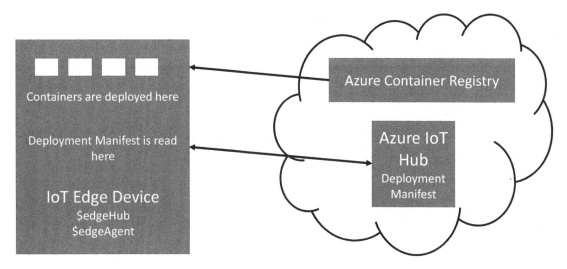

Figure 3-7. *Key modules in container deployment at the edge*

In the spring of 2019, Microsoft began to preview Azure SQL Database Edge, a variation of the Azure SQL Database designed to run on ARM-based or x64-based edge devices that are configured with only about 1 GB of memory. The installation procedure for the database onto the device is through containers.

Azure SQL Database Edge has a streaming engine built on Azure Stream Analytics that enables queries to connect through the Azure IoT Hub or Azure Event Hub on to backend Azure cloud services. Bidirectional data movement is supported. The database can be utilized in disconnected as well as cloud-connected scenarios.

For local machine learning applied to data in the edge device, both R and Python are supported through external procedures. The predictive functionality provided by Azure SQL Database Edge is the same as that in SQL Server. There is support for processing and storing time series, graph, and JSON data. The database can be accessed using popular business intelligence tools such as Microsoft's Power BI.

Azure IoT Edge and Device Security

Microsoft defines an Azure IoT Edge security framework that provides a foundation for secure deployment of devices. Anchoring the framework is secure silicon that is a tamper-resistant root of trust. At middle layers of the foundation, highly assured booting of the device and secured execution environments relying on proper authentication, authorization, and attestation become critical. Runtime integrity monitoring of applications completes the foundation.

Table 3-1 illustrates how these principles are realized in IoT Edge devices.

Table 3-1. *Realizing security principles in IoT Edge devices*

Foundation Principles	How the Principles are Realized
Application runtime integrity monitoring	Azure IoT Edge
Systems resource access control and privileged actions	Secured operating system on device
High assurance bootstrapping and resiliency	Azure IoT Edge Security Manager
Tamper resistant root of trust/secure silicon	Secure silicon/device provided by manufacturer

At the root, a best practice is to specify devices that meet minimal physical requirements. Physical features, such as USB ports, are to be avoided if they are considered optional as they can expose the device to attack. Devices might also be protected from physical tampering through secure enclosures and other means. A Software Guard Extension (SGX) in device processors can assure normal processes are in enclaves that can't be overridden.

Microsoft's Azure IoT Edge Security Manager is provided with the Azure IoT Edge software. It provides a level of security for the device even if the operating system running on the device is compromised. Specifically, it is responsible for

- Secured and measured bootstrapping of the device

- Device identity provisioning and transition of trust

- Hosting and protection of the Device Provisioning Service

- Securely provisioning IoT Edge modules with unique identities

- Serving as a gatekeeper to device hardware root of trust

- Monitoring the integrity of IoT Edge operations at runtime

Device manufacturers are responsible for providing Trusted Platform Module (TPM) drivers, the TPM itself, and any custom hardware security modules (HSMs) and drivers. The Trust Computing Group provides specification of a Device Identifier Composition Engine (DICE) that can be used for creating cryptographic representations of device identities.

Further protection is provided by securing the operating system on the device through systems resource access control and establishing privileged actions. Runtime integrity monitoring provided by the Azure IoT Edge software helps protect the general computing environment for the device.

Authentication is used in the foundation to assure that only trusted parties, modules, and devices have access. Certificate-based authentication is derived from standards governing public key infrastructure (PKi) by the Internet Engineering Task Force (IETF) and is the primary means of authentication. Where devices or components do not support certificate-based authentication, extensibility in the security framework can be utilized to provide needed authentication.

Authorization refers to the permission scope that actors, modules, and devices are granted. It is usually configured at a least privilege level that provides just enough access to the resources and data needed to deliver the designed business solution. Authorization can be managed through certificate signing rights or role-based access control (RBAC) for some scenarios.

Attestation assures the integrity of software at boot-up of the device, during runtime, and during software updates. During secure boot-up, integrity of all software on the device is assured including the operating system, the runtimes present, and the configuration information. Runtime attestation detects malware injections, improper physical access, and improper configuration changes, with countermeasures provided by the device and security framework to combat these threats. Software attestation assures secure software patching and updates through measured and signed packages.

A device can be considered as trusted by meeting standards such as ISO/IEC 11889 that specifies the architecture, data structures, command interface, and behavior of a TPM. A TPM device is trustworthy for storage, measurement, and reporting. Trust in these three elements is assured through authorization by using certificates and through attestation, thus providing evidence of the accuracy of information. The platform also offers protected locations for keys and data objects and can provide integrity measurements of platform state.

You likely will also consider protecting devices from external threats that could be initiated using cloud resources. For example, Arm TrustZones could be established to secure boundaries between edge devices or IoT gateways and the cloud.

Note   One way to monitor transmissions from the devices is through a Security Information and Event Management (SIEM) tool. Microsoft began to preview its Sentinel tool as this book was being published and promoted its utilization of AI to provide an early warning of unusual events, including in IoT devices. In addition, Azure Security Center can be used to help find missing security configurations in IoT devices.

The Azure IoT Device Catalog

To better understand which devices can most easily be implemented in the Microsoft IoT architecture and enable the device manufacturers to verify this, the Azure IoT certification service (AICS) was created. Once certified, the devices are listed in the Azure IoT device catalog (`https://catalog.azureiotsolutions.com/`).

Some of the equipment listed in the catalog are build-your-own-device Microsoft Azure IoT Starter Kits. These kits typically include a breadboard, some sensors, LEDs, resistors, jumper wires, and other parts needed.

Figure 3-8 illustrates some certified devices and starter kits on a page presented in the Azure IoT device catalog web site.

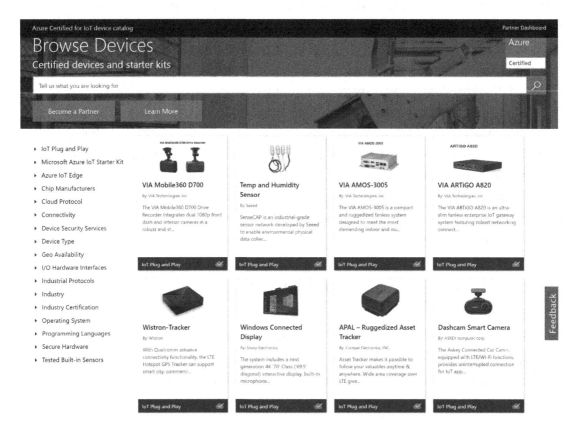

Figure 3-8. *Azure IoT device catalog certified devices and starter kits page*

To become certified, device partners first create a company profile and indicate the devices they wish to certify. For each device, the type of operating system for the platform and programming languages supported are provided in order to receive a certification test that can be run on the device. The vendor runs compatibility tests and provides packaging for installing agent on device as well as an example of using device with Azure IoT.

Note   A wide variety of operating systems can be run on the devices that are tested for certification. At the time this book was published, versions of Raspbian-stretch, CentOS, Debian, RHEL, Ubuntu, Ubuntu Server, Wind River, Windows 10 IoT Core, Windows 10 IoT Enterprise, Windows 10 Server, and Yocto had certification tests available.

Since 2018, Microsoft also provides a certification test for support of the Azure IoT Edge software on devices. During testing, IoT Hub primitives such as device-to-cloud, cloud-to-device, direct method, and device twin properties are validated, as is the presence of the EdgeAgent module on the device. A test is also run to ensure that a sample Edge module is successfully deployed to the device.

Optionally, security at the following four levels can be evaluated:

- **Level 1.** Custom security

- **Level 2.** Azure Device SDK

- **Level 3.** Azure Device SDK, FIPS 140-2 Level 2, and Common Criteria EAL 3+

- **Level 4.** Azure Device SDK, FIPS 140-2 Level 3, and Common Criteria EAL 4+

Level 2 can be reached through validation of base security processes at the Edge and all transactions monitored in accordance with a deployment risk assessment with no secure hardware in place. Level 4 can be reached by providing a secure element that includes a stand-alone security processor, secure hardware protection of storage, session key generation, authentication, and certificate processing in place. Level 4 can also be reached by providing a secure enclave containing an integrated security processor, providing secure element features, featuring protection of the execution environment, and providing metering, billing, secure I/O, and secure logging.

In 2019, Microsoft introduced IoT Plug and Play, an open modeling language to connect IoT devices to the cloud without having to write embedded code. IoT Plug and Play devices are defined by a device capability model in a JSON-LD document. Device properties including attributes (such as firmware version), device settings, telemetry sensor readings and alerting events, and available device commands are described in the model.

Devices listed in the Azure Certified for IoT Device Catalog are described by the following capabilities and properties:

- **IoT Plug and Play.** Certified for this capability?

- **Microsoft Azure IoT Starter Kit.** Yes or no?

- **Azure IoT Edge.** Certified for this capability?

- **Chip Manufacturer.** Intel, Microchip, Espressif, Qualcomm, Broadcom, Texas Instruments, NXP/Freescale, Nvidia, STMicroelectronics, VIA Technologies, or other.

- **Cloud Protocol.** Supports AMQPS, AMQPS over WebSockets, MQTT, MQTT over WebSockets, and/or HTTPS.

- **Connectivity.** Bluetooth, LAN, WIFI, LTE, 3G, and/or other.

- **Device Security Services.** Managed PKI (CSR, CRL, etc.), Symmetric Key Provisioning, firmware update and integrity, secure hardware attestation, secure hardware disablement, authentication and data protection, identity management, device management, and/or others.

- **Device type.** Gateway, industry tablet, mobile POS, or other.

- **Geo availability.** Worldwide, Europe, APAC, America, and/or Africa.

- **I/O hardware interfaces.** GPIO, I2C/SPI, COM (RS232, RS485, RS422), USB, and/or others.

- **Industrial protocols.** CAN bus, EtherCAT, Modbus, OPC Classic, OPC UA, PROFINET, ZigBee, PPMP, and/or others.

- **Industry.** CityNext (Smart Cities), discrete manufacturing, government, health, hospitality, insurance, media and cable, power and utilities, process manufacturing, retail and consumer goods, banking and capital markets, or others.

- **Industry certification.** Yes or no.

- **Operating system.** Windows IoT Core, Windows IoT Enterprise, Windows 8/10, Debian, Arduino, Windows Server, Ubuntu, iOS, Mbed, Yocto, RTOS, Fedora, Android, Raspbian, RHEL CentOS, Wind River, no OS, or others.

- **Programming languages.** C, C#, Java, JavaScript (node), and/or Python.

- **Secure hardware.** TPM, DICE, SIM and eSIM, Smartcard, and/or others.

- **Tested built-in sensors.** GPS, touch, LED, light, gas, noise, proximity, temperature, humidity, pressure, accelerometers, weight, soil alkalinity, vibrations, image capture, motion detection, chemical/compound presence, no built-in sensors, and/or others.

Note   The first starter kits for Azure Sphere microcontroller units (MCUs) providing a hardware-based root of trust were being released as this book was being published. For example, the Avnet Azure Sphere Starter Kit contains a MediaTek MT3620 MCU and includes a three-axis accelerometer, three-axis gyro, temperature sensor, and ambient light sensor.

The MCU features the Microsoft Pluton security subsystem, a full memory management unit for compartmentalization of processes, real-time operating system, Wi-Fi radio, multiplexed I/O, hardware firewall, and integrated RAM and flash memory. The software stack features a security monitor protecting the hardware and custom Linux kernel. Operating system services include a device authentication client, over-the-air (OTA) update client, application management, and networking management. Custom applications are typically developed in C using Visual Studio.

When looking for devices or gateways, you can search the catalog using these parameters. For example, you might want to gather and transmit data from computer numerical control (CNC) machinery on your factory floor that automates control of machining tools used in milling, turning, and grinding raw materials. You can search for IoT gateways in the catalog that support industrial protocols such as ABB, FANUC, or Modbus, and you will see the vendors of such equipment present in the catalog.

Throughout this chapter, we've focused on edge devices and Azure IoT Edge software but have frequently mentioned the critical role of the Azure IoT Hub in communicating and managing these devices. We next will look at the Azure IoT Hub in more detail in Chapter 4.

CHAPTER 4

Azure IoT Hub

IoT edge devices capture data gathered in events and transmit lightweight notifications of conditions or discrete state changes. Condition changes are commonly reported in a time series of interrelated events that are then analyzed to determine what has happened. Discrete events indicating a change in state can drive the need to perform specific actions.

When thousands or more IoT devices are deployed, thousands to millions of events per second can land in the cloud for further data processing and analysis. Microsoft's first hub capable of ingesting telemetry from large numbers of IoT devices at rates of over 1 GB per second was its Azure Event Hub cloud service in the Microsoft Azure Service Bus. The Event Hub became generally available in November 2014. It supports AMQP, AMQP over WebSockets, and HTTPS as protocols for communications to the cloud. However, though Event Hubs were designed for streaming data ingestion, they were not designed to enable communications back to IoT devices.

In February 2016, Microsoft announced general availability of the Azure IoT Hub, a cloud service designed for both IoT device-to-cloud and cloud-to-IoT device communications. Building upon previous Event Hub capabilities for ingestion, the IoT Hub also supports MQTT and MQTT over WebSockets protocols, per-device identity, file upload from devices, device provisioning, device twin and device management, device streams, and the Azure IoT Edge. Through the IoT Hub, it is possible to track device creation, device connections, and device failures. Additional communications protocols can be supported through deployment of custom Azure IoT protocol gateways in the cloud, though protocol conversion is more often deployed in the IoT Edge (as described in the previous chapter).

© Robert Stackowiak 2019
R. Stackowiak, *Azure Internet of Things Revealed*, https://doi.org/10.1007/978-1-4842-5470-7_4

Today, Microsoft recommends deployment of the IoT Hub for all scenarios where IoT devices are connected to the Azure cloud. We focus this chapter on the following topics:

- IoT Hub capabilities

- Configuring the IoT Hub

- IoT Hub Performance Monitoring

- IoT Hub Device Provisioning

- IoT Hub Availability and Disaster Recovery

IoT Hub Capabilities

The Azure IoT Hub provides a cloud landing spot for telemetry gathered by IoT devices. It also has a central role in configuring and controlling devices by

- Storing, synchronizing, and enabling querying of device metadata and state information using device twins

- Providing the ability to set the device state either by individual device or based on common characteristics of multiple devices

- Providing automatic response to device-reported state changes through message routing integration

The status of the IoT Hub is determined through monitoring of device identity operations, device telemetry and diagnostics, cloud-to-device commands, and connections. Each device uses its own security key to connect to the IoT Hub. You can individually whitelist or blacklist each device providing complete control over device access.

Device applications can read and receive notification of changes in desired properties that reside in the device twin in the IoT Hub. The desired properties are modifiable by the IoT solution backend. Reported properties in the device twin are modifiable by device applications, and changes are read by the backend. Tags stored in the device twin contain device metadata and are accessible by only the backend. In addition to storing device metadata and reporting on current state information, device twins can be used to synchronize the state of long-running workflows between devices and backend applications.

The IoT Hub can also be thought of as the front door to several key Microsoft backend cloud-based services. Some of the services include

- Azure Stream Analytics

- Azure Time Series Insights

- Azure Databricks

- Apache Spark

- Apache Storm (spout)

- Azure Functions

- Azure Logic Apps

Azure Functions enable creation and deployment of actions that contain custom written code in C#, F#, or Java. Azure Logic Apps provide a collection of pre-defined actions that can be orchestrated using a GUI-based development environment. Both are deployed as serverless workloads. We'll describe the analytics, machine learning, and other related backend capabilities and components in Chapter 5.

Data can be retained in the IoT Hub's built-in Event Hubs for up to 7 days. Messages that are at maximum message sizes are retained for 24 hours at a minimum.

Configuring the IoT Hub

An Azure IoT Hub can be created and managed in a variety of ways including through the Azure Portal, Azure CLI, or using PowerShell. Here, we'll describe creating an IoT Hub using the Azure Portal.

As shown in Figure 4-1, one begins by assigning an Azure subscription, creating a new or using an existing Azure Resource Group, choosing an Azure Region for deployment, and providing a name for your IoT Hub.

Figure 4-1. *Initial IoT Hub resource assignment and naming*

The next step is to choose the pricing and scaling tier (basic or standard levels for production) based on the features desired and number of messages per day you want the IoT Hub to be capable of handling. Pricing in levels is provided on a per-month basis in the portal interface.

Both basic and standard options support similar numbers of messages per unit per day as well as the maximum units that can be assigned. The option levels are shown in Table 4-1. At the standard level, cloud-to-device command enablement, IoT Edge support, and device management are also provided (whereas these features are missing in the basic levels). Basic levels can be upgraded to standard levels.

Table 4-1. *Basic and standard levels message scalability*

Option Level	Messages/ Unit/Day	Number of Units
Basic B1	400 K	1 to 200
Basic B2	6 M	1 to 200
Basic B3	300 M	1 to 10
Standard S1	400 K	1 to 200
Standard S2	6 M	1 to 200
Standard S3	300 M	1 to 10

During this step, you also choose the number of IoT Hub units to be deployed in the pricing tier that you selected. An appropriate number of units is selected that will deliver the desired number of messages per day by using a slider bar in the interface.

The portal interface for selection of the pricing and scaling tier and number of IoT Hub units is illustrated in Figure 4-2.

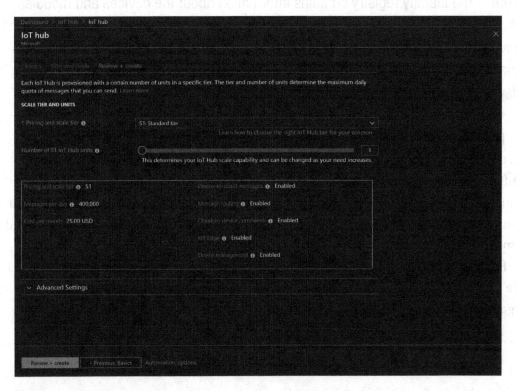

Figure 4-2. *IoT Hub pricing tier and scale selection*

Once you've completed this step, you can review your selections and create the IoT Hub. Resource capabilities that you've defined are then provisioned.

Managing the IoT Hub

When a hub is created, you can review parameters associated with the IoT Hub through the overview provided in the Azure Portal. You can also view the hub's activity log of operations and adjust IoT Hub access control and settings.

For example, in settings, you can change the pricing and scale tier and number of units if the projected number of messages being handled changes. You can also adjust operations monitoring (such as turning event logging on or off) and specify valid IP address ranges that the IoT Hub will accept. Shared IoT Hub access policies that can be adjusted in settings include permissions for identity registry reads, identity registry writes, the service connect to service endpoints, and the device connect sending and receiving of messages.

Note   The identity registry contains information about the devices and modules allowed to connect to the IoT Hub and stores credentials used in authenticating the devices and modules. You must add device IDs and keys to the identity registry to enable the devices to connect to the IoT Hub, normally through the Azure IoT Hub Device Provisioning Service.

The portal also provides access to explorers for query and to IoT devices. You can use the portal to choose automatic device management (IoT Edge and IoT device configuration), designate file uploads and message routing, manually initiate failover from an IoT Hub primary to secondary location, access the Azure Security Center, and monitor the hub for alerts and metrics. Finally, you can determine resource health and/or make a support request.

Figure 4-3 illustrates a portion of the options that you have available on the left side of this IoT Hub view in the portal. On the right, overview charts indicate recent ongoing activity.

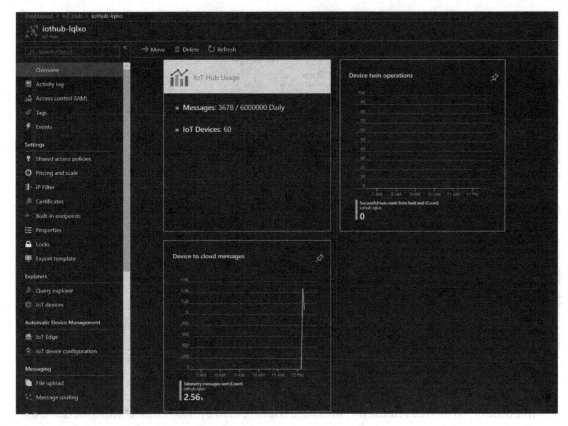

Figure 4-3. *Overview of IoT Hub activity*

Message Routing and Event Routing

Using the portal, you can leverage custom endpoints when you create, define, and manage routing of messages from the Azure cloud to devices. The maximum message size supported is 256 KB. Near real-time messages are received at endpoints in the order in which they are sent. Up to 10 custom endpoints and 100 routes can be created per IoT Hub. As illustrated in Figure 4-4, the custom endpoints can be added as Event Hubs, a Service Bus queue, Service Bus topics, and Blob storage.

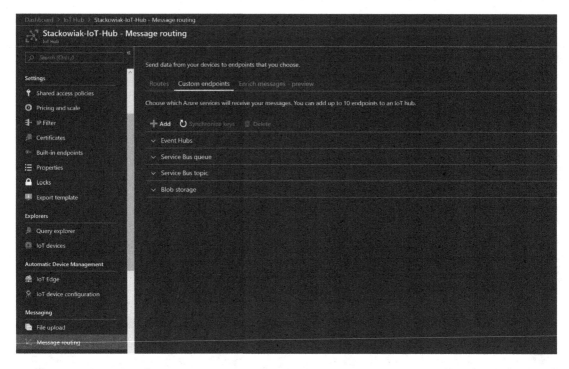

Figure 4-4. *IoT Hub message routing options*

In addition to device telemetry, message routing also enables the sending of device twin change events and device life cycle events. Events published by the IoT Hub include

- Device registration to an IoT Hub.

- Device deletion from an IoT Hub.

- Device connection to an IoT Hub.

- Device disconnection from an IoT Hub.

- Device telemetry message is sent to an IoT Hub.

You can query message application and system properties, message bodies, device twin tags, and device twin properties.

Alternatively, you might want to set up a publish-and-subscribe event routing service by integrating the Azure IoT Hub with an Azure Event Grid. In this configuration, the IoT Hub publishes events to endpoint subscribers. Maximum message size is also 256 KB in this scenario. You can filter data using message properties, message body, and device twin in the IoT Hub before publishing to the Azure Event Grid.

It is important to realize that events will not necessarily arrive at endpoints in the order in which they were published. In such scenarios where order is important, message routing should be chosen over leveraging an Event Grid.

An Azure IoT Hub can support up to 500 endpoints when integrated with Azure Event Grid. Endpoint types supported include Azure Automation, Azure Functions, Azure Event Hubs, Azure Logic Apps, Storage Blobs, Custom Topics, Queue Storage, Microsoft Flow, and third-party services through WebHooks.

IoT Hub Performance Monitoring

Through the Azure Portal, you can monitor performance of your IoT Hub. Typical statistics presented include the following:

- Active devices

- Total devices

- Total messages

- Messages per second

- Failed messages

- Failed device connections

- Failed twin updates

If you are going to deploy new or untested devices and want to understand what monitoring output from them might look like, you can choose to simulate output from the devices during solution development. Microsoft provides a device simulation solution accelerator for this purpose. It is found on the Microsoft IoT solution accelerator web site at `https://azure.microsoft.com/features/iot-accelerators`.

The sample simulations are easily spun up, and the framework provided can be used to create custom and/or more advanced simulated devices. Figure 4-5 illustrates sources of data in a sample simulation of delivery trucks that will send messages to your IoT Hub.

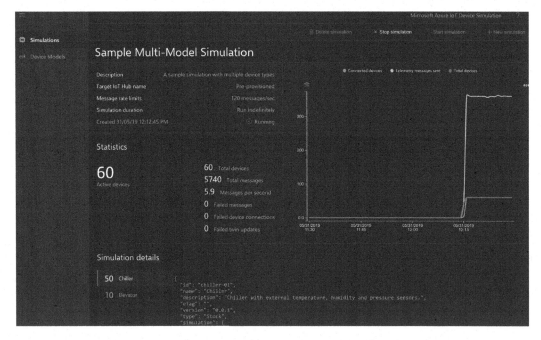

Figure 4-5. *Message sources in the Microsoft IoT device simulation accelerator*

Sample output from a simulation is presented in Figure 4-6. We see an indication as to when the simulation began and the other performance metrics that are gathered.

Figure 4-6. *Sample output from multi-model simulation*

IoT Hub Device Provisioning

The IoT Hub Device Provisioning Service is used to provision IoT devices without the presence of hardcoded IoT connection information in the device. During manufacturing of the device, the device is programmed to call the Provisioning Service when it is turned on so that it can get connection information and its IoT solution assignment.

The Provisioning Service is used to enable the following tasks:

- Load balancing of devices across multiple hubs.

- Connect devices to IoT solutions based on transaction data or use case.

- Connect devices to IoT Hubs with lowest latency using AMQP, AMQP over WebSockets, MQTT, MQTT over WebSockets, or HTTPS.

- Reprovision the device when the device is changed.

- Roll keys used by the device to connect to the IoT Hub (if not using X.509 certificates to connect).

Features present in the Device Provisioning Service include

- Secure attestation support for X.509 and TPM identities

- An enrollment list that includes devices or device groups that might register and device configuration information

- Allocation policies that control how the Device Provisioning Service assigns devices to IoT Hubs

- Monitoring and diagnostics logging

- Multiple IoT Hub assignments for devices

- Cross-region assignments of IoT Hubs for devices

- For service operations, uses HTTPS as a protocol

Note   The Device Provisioning Service can also be used to pre-authorize devices paired with over-the-air software update solutions such as Mender.io. Such solutions can help assure that software updates are performed in a secure manner and that software on the devices is always current.

After an IoT Hub is created, an IoT Hub Provisioning Service is set up using the Azure Portal by providing a name for the Device Provisioning Service, choosing the subscription you want to assign the Provisioning Service to, creating or assigning a resource group to the new Provisioning Service, and selecting a location close to your device. The device manufacturer adds device registration information to the enrollment list.

A series of automated steps follow to establish the connection between the device and the IoT Hub and begin provisioning. The device first contacts the Provisioning Service endpoint that was set at the factory and passes its identification information to prove its identity. The Provisioning Service validates the device's registration ID and key against the enrollment list either using a Trusted Platform Module (TPM) nonce challenge or X.509 for verification. The Provisioning Service then registers the device with an IoT Hub and populates the device's desired twin state. After the IoT Hub returns device ID information to the Provisioning Service, the Provisioning Service returns IoT Hub connection information to the device and the device connects to the IoT Hub using one of the supported cloud protocols. The device then gets the desired state information from its device twin in the IoT Hub.

You can assign another device to a different IoT Hub in a similar manner and then add an enrollment entry for the second device. The allocation policy selected determines how devices are assigned to IoT Hubs. Available allocation policies include

- **Lowest Latency.** Devices are provisioned to the IoT Hub with the lowest latency to the device.

- **Even Weighted Distribution.** Linked IoT Hubs are equally likely to have devices provisioned to them (by default).

- **Static Configuration using the Enrollment List.** Specification of the desired IoT Hub in the enrollment list takes priority.

The Device Provisioning Service would then be linked to the second IoT Hub.

IoT Hub Availability and Disaster Recovery

Within a region, the IoT Hub service provides high availability through redundancies in nearly all service layers. Microsoft's Service Level Agreement (SLA) for the Azure IoT Hub is 99.9 percent availability during which the Hub can send messages to and from registered devices. During this time, the service can perform create, read, update, and delete operations on the IoT Hubs. Similarly, Microsoft states that 99.9 percent of the time, provisioning service will be able to receive provisioning requests from devices and register them to an IoT Hub.

That said, device applications should have retry policies and procedures built in to deal with situations caused by transient problems. Examples of these situations include

- Fixing dropped network connections

- Switching between network connections

- Reconnecting after transient connection errors

To reach a higher level of availability, a disaster recovery plan can be put into place to account for the extremely rare instance in which a data center becomes unavailable. To implement a regional failover model, you must have a secondary IoT Hub and device routing logic in place. The secondary IoT Hub must contain all device identities through replication from the primary IoT Hub. When the primary region becomes available again, all state and data created at the secondary site must be migrated back to the primary site.

In this scenario, recovery point objectives (RPOs) are 0 to 5 minutes of data loss for identity registry, device twin data, cloud-to-device messages, and parent and device jobs. All unread messages are lost for device-to-cloud messages, operations monitoring messages, and cloud-to-device feedback messages. The recovery time objective (RTO) where manual failover is put into place using the Azure Portal ranges from 10 minutes to 2 hours. For Microsoft-initiated failovers, RTO ranges from 2 to 26 hours.

Once you've set up your IoT Hub(s) and start receiving data from your devices, you are likely ready to analyze the data that is being gathered. That is the subject of the next chapter.

CHAPTER 5

Analyzing and Visualizing Data in Azure

In this chapter, we explore processing, analyzing, and visualizing data that lands in the Azure cloud at a deeper level than in the previous introduction provided in Chapter 2. Our goal is to help you understand how each of the platforms and tools described are best utilized as you consider their inclusion into your own architecture. You should gain insight into where and how to deploy them.

As data coming from IoT devices is most often semi-structured, we focus the data management discussion in this chapter on Azure HDInsight and Cosmos DB. Data warehouses are also often part of the architecture as they enable business intelligence and analytics solutions where the data lines up neatly into rows and columns. We'll address how they and associated tools can fit into this architecture in Chapter 7 when we consider integration with legacy data solutions.

The following components in the IoT architecture will be covered here:

- Azure Stream Analytics

- Time Series Insights

- Azure Databricks

- Semi-structured Data Management (Azure HDInsight and Cosmos DB)

- Azure Machine Learning

- Cognitive Services

- Data Visualization and Power BI

- Azure Bot Service and Bot Framework

© Robert Stackowiak 2019
R. Stackowiak, *Azure Internet of Things Revealed*, https://doi.org/10.1007/978-1-4842-5470-7_5

Azure Stream Analytics

The Azure Stream Analytics in-memory streaming data analytics and event processing engine is designed to run transformation queries against input coming from IoT Hubs, Event Hubs, and Azure Blob Storage. It can be deployed in Azure or at the edge in containers deployed to devices.

Transformation queries are based on SQL and are used for filtering, sorting, aggregating, and joining streaming data or applying geospatial functions. You can also define function calls to the Azure Machine Learning service and/or create user-defined JavaScript or C# functions that you run in jobs. Stream Analytics jobs can be created using the Azure Portal, Azure PowerShell or Visual Studio.

Stream Analytics can process millions of events every second in Azure. Through partitioning, complex queries can be parallelized and executed on multiple nodes. The Stream Analytics SLA guarantees 99.9 percent availability for event processing every minute. There are built-in checkpoints and recoverability if delivery of an event fails.

Output can be sent to a monitored queue (such as to an Azure Service Bus, Azure Functions, or Azure Event Hubs) to trigger alerts or custom workflows. Data can be stored in downstream Azure data management solutions such as Azure Data Lake Storage, Cosmos DB, SQL Database, or SQL Data Warehouse and is often visualized in Power BI.

When you create a new job using the Azure Portal, you begin by defining the job name, choose a subscription and resource group to use, choose a location, and indicate the hosting environment and (in cloud deployment) the number of streaming units that provide a pool of computation resources.

You can then set inputs and outputs and define a query stream using the Azure Portal interface pictured in Figure 5-1. You also start, stop, and monitor jobs through this interface.

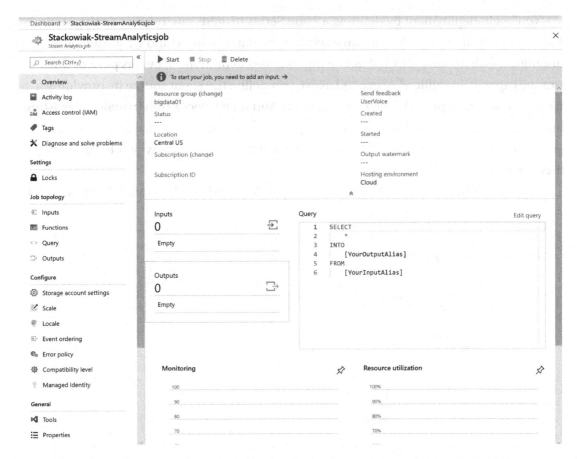

Figure 5-1. *Defining inputs, outputs, and queries in a Stream Analytics job*

Streaming inputs can be defined coming from IoT Hubs, Event Hubs, or Blob Storage. Reference inputs can be defined coming from Blob Storage or a SQL Database. Outputs can be designated to Event Hubs, SQL Database, Blob Storage, Table storage, Service Bus topics, Service Bus queues, Cosmos DB, Power BI, Azure Data Lake Storage, or Azure Functions.

Time Series Insights

IoT devices commonly send telemetry messages to the cloud in a time series (i.e., the data is timestamped). The data initially lands in Azure in the Azure IoT Hub or Azure Event Hub. Time Series Insights connects to Azure IoT Hubs and Azure Event Hubs and parses JSON from these incoming messages. Metadata is joined with telemetry, and the data is indexed in a columnar store. The data is stored in memory and SSDs for up to 400 days. It can be queried using the Time Series Insights explorer or using APIs in custom applications.

You begin deployment by defining a Time Series Insights environment to be used. The Azure Portal prompts you for an environment name, subscription, location, and pricing tier (where tiers selected define ingress rates in millions of events per day and storage capacity in millions of events). Next, you define the event source by providing a name and source type (IoT Hub or Event Hub). You then select a hub (usually an existing hub) and apply an IoT Hub access policy name. For IoT Hubs, you also set a consumer group parameter and can create an event source timestamp property name. You can then create the Time Series Insights environment.

Figure 5-2 illustrates a summary of these selections (with subscription id not visible).

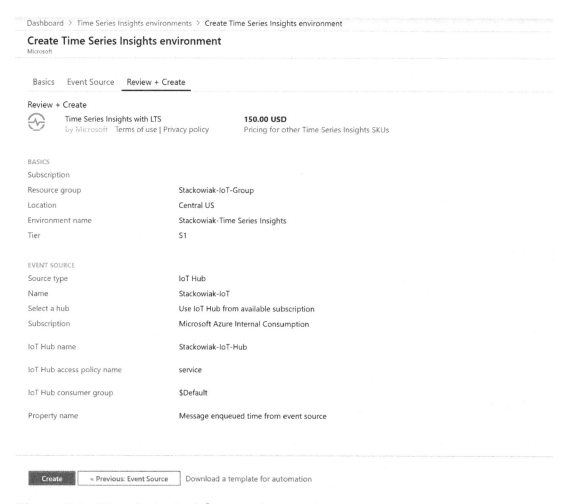

Figure 5-2. *Time Series Insights creation summary*

Once operational, you can query data using the Time Series Insights explorer or through APIs. Figure 5-3 shows one of the visualizations provided by the Time Series Insights explorer. Here, the tool is being applied to sample data provided by Microsoft that helps explain explorer functionality.

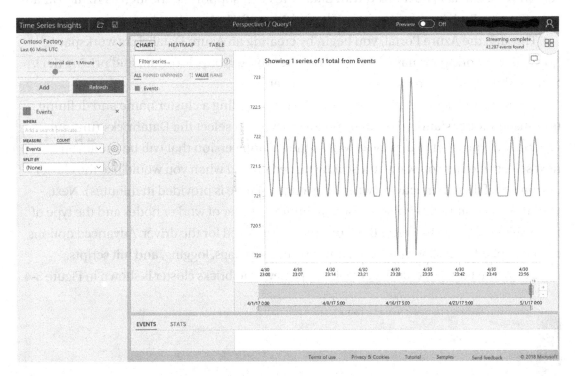

Figure 5-3. *Time Series Insights explorer*

Time series data can be monitored to determine the health of the device. You can apply perspective views and discern patterns when performing root cause analysis. Azure Stream Analytics might also be inserted into the data flow to help you find anomalies and send alerts.

Azure Databricks

Azure Databricks enables a fully managed Apache Spark cluster in the cloud. You can program in Python, R, Scala, SQL, and Java and utilize the Spark Core API. As the entire Spark ecosystem is provided, you can use Spark SQL to work with tabular data stored in DataFrames, process and analyze streaming data in real-time (with integration to HDFS,

Flume, and Kafka), utilize GraphX, and access the MLib machine learning library that includes classification, regression, clustering, collaborative filtering, and dimensionality reduction algorithms.

The Databricks Runtime is built upon this Spark base and can be deployed as serverless. It can also be utilized with datastores that support Spark such as Azure Data Lake Storage, Blob Storage, Cosmos DB, and Azure SQL Data Warehouse.

Through the Azure Portal, you begin by creating an Azure Databricks workspace (providing a workspace name, subscription, resource group, location, and pricing tier). You are then ready to create a Databricks cluster.

Databricks cluster creation begins with you providing a cluster name and defining the cluster mode (standard or high concurrency). You select the Databricks runtime version that you wish to deploy as well as the Python version that will be used. You next select whether you want autoscaling turned on and when you would like the cluster terminated if there is inactivity (where the length of time is provided in minutes). Next, you select the minimum number and maximum number of worker nodes and the type of hardware used. You also select the type of hardware used for the driver. Advanced options can be applied including Spark configuration options, tags, logging, and init scripts.

The Azure Portal interface for creation of a new Databricks cluster is shown in Figure 5-4.

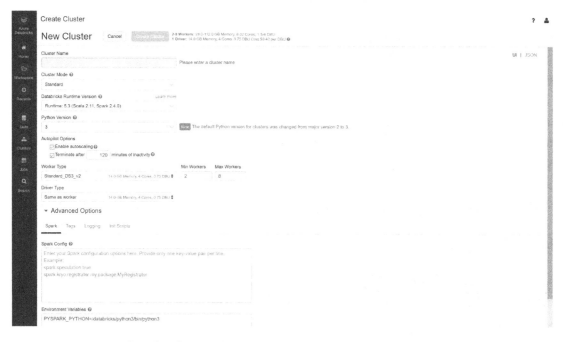

Figure 5-4. *Databricks cluster creation in the Azure Portal*

Once you've created the cluster, Databricks will present a screen like the following shown in Figure 5-5. You will see your resource group, the name of the managed resource group, the subscription information, the URL for Azure Databricks at the location you selected, and the pricing tier. From this screen, you can launch the workspace. You can also follow links to documentation, getting started, importing data from a file, importing data from Azure storage, access to a notebook, and the Administrators' Guide.

Figure 5-5. *An initial view of Databricks after cluster creation*

Upon launching the workspace, you will be logged in using your Azure Active Directory identity. Your Databricks workspace will appear like that shown in Figure 5-6. Common tasks you will likely want to execute are shown on the left in the figure including creating a new notebook (through a web-based application that enables creating and sharing of documents that contain the live code, equations, visualization, and descriptive text); uploading data (from a file, DBFS, Azure Blob Storage, Azure Data Lake Storage, Cassandra, JDBC, Kafka, Redis, or Elasticsearch); creating a table; creating a new cluster, new job, or new MLflow experiment; importing a library; or reading the documentation. As you create notebooks, they will appear under "Recents" heading.

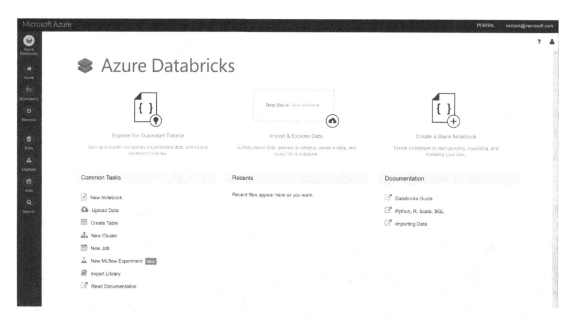

Figure 5-6. *Databricks workspace*

A Quickstart Notebook is provided by Microsoft as an example. A portion of that notebook is shown in Figure 5-7.

Figure 5-7. *Typical notebook view in Databricks*

Within notebooks, you can provide code in R, Python, Scala, or SQL and provide supporting commentary and documentation. You can visualize data using tools such as Matplotlib, ggplot, or d3. Power BI provides additional data visualization capabilities as described later in this chapter.

Semi-structured Data Management

In addition to processing and analyzing data at the edge or within the data stream, machine learning models are often developed through analysis of historical data over lengthy time periods. Such data needs to land in a data management system designed for storing and analyzing such data.

NoSQL databases are ideal for semi-structured data. At the beginning of this century, Hadoop established itself as a popular open-source historical data store. The Hadoop version available in a PaaS offering from Microsoft is Azure HDInsight. More recently, NoSQL databases that are globally distributed have proven their ability to scale to enormous sizes. Microsoft's PaaS offering here is Cosmos DB.

In this section of the chapter, we'll describe Azure HDInsight and Cosmos DB. Either can be created through the Azure Portal, Azure CLI, and PowerShell. We'll describe the creation of these data management systems using the Azure Portal.

Azure HDInsight

Azure HDInsight is Microsoft's cloud-based offering that consists of Apache Hadoop components in the Hortonworks Data Platform (HDP). HDInsight clusters enable deployment of Hadoop, Spark for in-memory processing, Hive low-latency analytical processing (LLAP) for queries, Kafka and Storm for processing streaming data, HBase (a NoSQL database), and/or ML Services.

Clusters are monitored using Apache Ambari and the Azure Monitor. Cluster health and availability, cluster resource utilization, performance across the entire cluster, and YARN job queues are monitored with Ambari. Resource utilization at the virtual machine level is monitored using Azure Monitor. Information about the workloads being run is present in the YARN resource manager and in Azure Monitor logs.

Languages native to Hadoop include Pig Latin, HiveQL, and SparkSQL. Programming languages supported include Java, Python, .NET, and Go. Other languages, such as Scala, can be deployed in Java Virtual Machines. Typical development environments that are used include Visual Studio, Visual Studio Code, Eclipse, and Intellij for Scala.

Microsoft released several versions of the distribution that was initially deployed to either Azure Data Lake Storage (ADLS) Gen1 featuring a hierarchical file system or to Blob Storage. The release of ADLS Gen2 provides a combination of hierarchical file system and Blob Storage capabilities, and it is now commonly selected for deployment of HDInsight clusters.

An Azure Blob System (ABFS) driver is provided with HDInsight, as well as Databricks, providing access to storage. If you are going to use Azure Data Lake Storage in the deployment, ADLS must be created first.

Note   Using the Azure Portal to create ADLS, you first select a subscription and resource group for the storage account, give it a name, and set the location. You can also specify performance, account kind, replication, and access tier. Next in advance, you can set security and virtual network fields (if not satisfied with the defaults provided). In the Data Lake Storage Gen2 section, you set the hierarchical namespace to enabled.

Deploying HDInsight is a three-step process using the Azure Portal. You begin by defining basic properties including a name for the Hadoop cluster, subscription to be used, cluster login name and password, secure shell (SSH) username, password for SSH, resource group for the cluster and dependent storage account, and location. You also select the cluster type and select the version of HDInsight that you want to deploy.

Next, you select the storage type (either Azure Blob Storage or Azure Data Lake Storage) and the storage account (from your subscriptions or from another subscription by providing an access key). You can choose to preserve metadata outside of the cluster by linking a SQL database for Hive and/or Oozie.

In the third step, you receive a summary of your selections and can edit those selections. When satisfied with the choices made, you next create the cluster. Clusters can take up to 20 minutes to be created.

A common means of moving data into and out of HDInsight when connected to the IoT Hub is to use Apache Kafka. You would begin by installing the IoT Hub Connector on an edge node in the HDInsight cluster. You would then get the IoT Hub connection information, configure the connector to serve as a sink and/or source for data movement, and start the connector.

Cosmos DB

Cosmos DB is a globally distributed multi-model database. The database can manage key-value, columnar, document, and graph data. Indexing of all data is automatic, and no schema or secondary indexes are required. Data can be made accessible using SQL, the MongoDB API, Cassandra API, Azure Table Storage API, or Gremlin API.

Storage and throughput are elastically scaled across regions making it possible to handle hundreds of millions of requests per second. Since the data is globally distributed, SLAs are provided where 99 percent of read and write requests will occur within 10 milliseconds in the region closest to the user. SLAs of 99.999 percent for high availability can also be attained.

Depending on performance needed, a variety of data consistency levels can be specified. The data consistency levels can be designated as follows:

- **Strong Consistency.** Only when an operation is complete is it is visible to all.

- **Bounded Staleness Consistency.** Read operations will lag writes based on consistent prefixes or time intervals; this level preserves 99.99 percent availability.

- **Session Consistency.** Consistent prefixes are applied with predictable consistency for a session, featuring high read throughput and low latency.

- **Consistent Prefix Consistency.** Reads will never see out-of-order writes.

- **Eventual Consistency.** Provides the lowest cost for reads; however, there is a potential for reads seeing out-of-order data.

When creating a Cosmos DB database using the Azure Portal, you provide basic information on the first Cosmos DB Account screen, then networking and tagging information, and finally review and creation of the Cosmos DB account. Figure 5-8 illustrates the first screen in the creation process in which you provide subscription information, the name of the resource group, an account name, specify the API that will be used and the originating location, and enable support of geo-redundancy and multiregion writes.

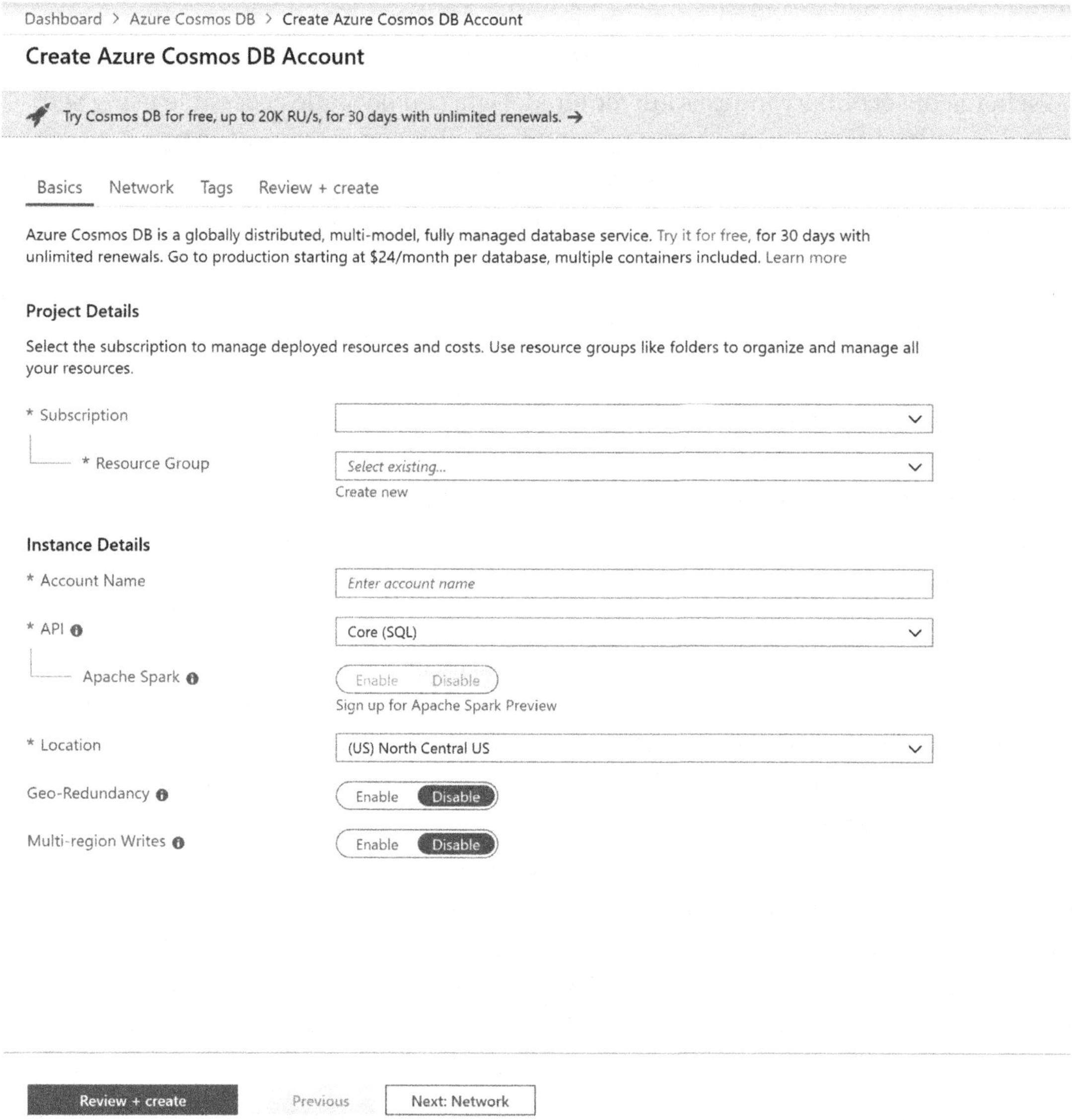

Figure 5-8. *Initial configuration of Cosmos DB*

Loading of data into Cosmos DB from IoT devices can programmatically take place in many ways. Some examples include

- Loading of data from the Databricks in-memory engine (where data initially landed in Azure in the IoT Hub and then was loaded into Databricks)

- Creating stored procedures and Logic Apps in an Event Grid deployed in the IoT Hub that write data into Cosmos DB

- Deploying Azure Functions in IoT Hub message routing that write data to Cosmos DB

Azure Machine Learning

Azure Databricks, previously described in this chapter, is only one of the means to build machine learning solutions in Azure. In this section, we'll look at the following:

- Azure Machine Learning Studio

- Azure Machine Learning service (including development environments)

Azure Machine Learning Studio

Azure Machine Learning Studio is an online development environment providing a drag-and-drop interface that is used in building, testing, and deploying predictive analytics solutions. At the time this book was published, experiments were limited to training sets of no more than 10 GB in size. However, a visual interface based on ML Studio integrated with the Azure Machine Learning service was in preview enabling preparing, training, and deployment with much larger datasets typically used by data scientists.

Drag–drop modules and functions are provided for building experiments that include saved datasets, trained models, transforms, data format conversions, data transformation, feature selection, machine learning, Open CV library modules, Python language modules, R Language Modules, statistical functions, text analytics, time series anomaly detection, and web services. The machine learning category includes functions used in evaluation, initializing the model using anomaly detection, classification, clustering, or regression algorithms, scoring, and training. Statistical functions include math operations, linear correlation, probability distribution functions, t-test, and descriptive statistics reporting.

Figure 5-9 shows the interface with icons representing projects, experiments, web services, notebooks, datasets, trained models, and settings on the far left, functions and modules to the right, then the canvas showing the experiment, and finally the properties and project information. Across the bottom, you have options to run history, save or save as the current experiment, discard changes, run the experiment, set up a web service, or publish to the ML Studio Gallery.

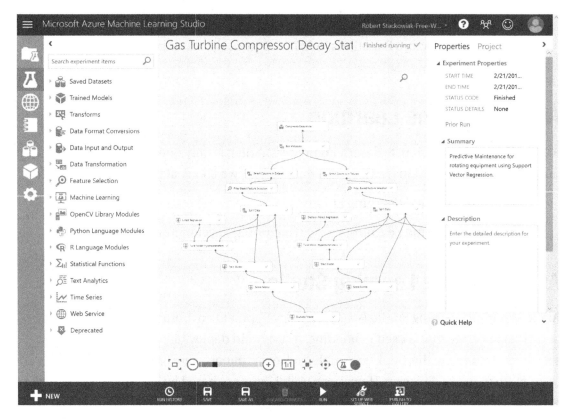

Figure 5-9. *Azure Machine Learning Studio experiment*

In the figure, we see a typical experiment data flow that begins with data input containing known outcomes, then preparing the data, splitting it for model training purposes, testing various mathematical models against the data, scoring them, and evaluating them for accuracy. Once we're satisfied with a specific model, we convert the training experiment into a predictive experiment and can deploy it as a web service. Sample code is also provided in C#, Python, and R.

Azure Machine Learning Service

The Azure Machine Learning service is Microsoft's PaaS offering used to train, deploy, and manage machine learning models at scales that data scientists typically work with. It is an open framework and can be used with open-source libraries that include MXNet, PyTorch, scikit-learn, and TensorFlow.

You begin by first generating a Machine Learning service workspace, typically through the Azure Portal. You provide a workspace name, subscription, resource group, and Azure region location for the workspace to be run.

In Figure 5-10, we see that a couple of Azure Machine Learning Workspaces have been created.

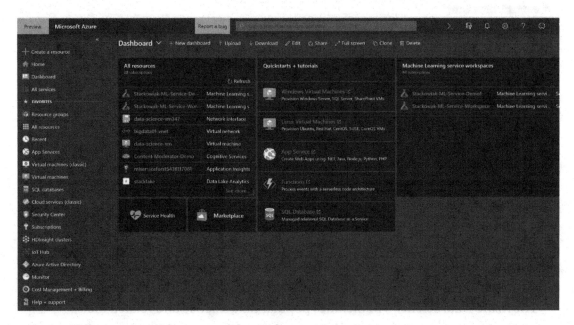

Figure 5-10. *Machine Learning services workspaces in the Azure Portal*

You'll have access to "Getting Started in Azure Notebooks," a Forum, samples in GitHub, and the documentation when you enter the workspace. You will also have access to other features under public preview.

Most data scientists prefer to write code (most often in Python) that performs data cleansing and transformation, simulation and modeling, machine learning, and data visualization. Jupyter Notebooks are open-source web applications that enable creating and sharing of documents that contain the live code, equations, visualization, and descriptive text. Azure Notebooks provide this capability, as Figure 5-11 illustrates.

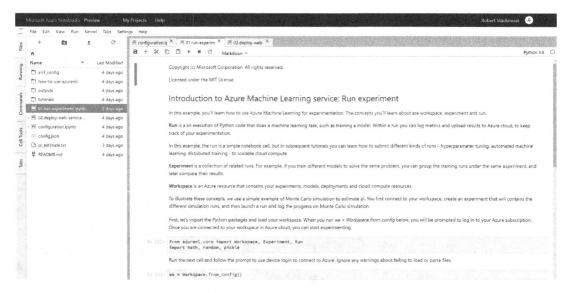

Figure 5-11. *Azure Notebook*

Azure Notebooks are a preinstalled free cloud service that support up to 4 GB of memory and 1 GB of data. To remove these limits, you can attach a Notebooks project to a VM running the Jupyter server or to the Azure Data Science Virtual Machine.

The Azure Data Science VM includes popular data science and related tools preinstalled and pre-configured and comes in Linux Ubuntu and Windows editions. Some of the tools that you will find here include Microsoft R/Open, Microsoft ML Server (with support for R and Python), Anaconda Python, various data management servers, Spark-based big data platforms used for development and testing, a Jupyter Notebook Server, IDE support for R Studio and Visual Studio, data movement and management tools, machine learning tools, and deep learning tools.

Microsoft developers will be happy to find that Visual Studio can also be used for building, testing, and deploying Azure Machine Learning service solutions. The code editor highlights syntax, provides intelligent code completion (known as Intellisense), and provides auto text formatting. You can debug your code locally by installing appropriate Python versions and libraries and the deep learning frameworks that you are using in your project.

Figure 5-12 illustrates Visual Studio being used in testing Python Code for Azure Machine Learning service, with Cloud Explorer shown on the left.

Figure 5-12. *Visual Studio and Python development for Azure ML service*

When you run your experiment, you can view the results through the Azure Portal interface into your workspace. You can apply active filters and view the maximum number of iterations to be run and the results of each iteration as shown in Figure 5-13. Above the experiment results in the figure, you also see tabs for pipelines, compute applied, models used, images (containers) created, deployments, and a summary of all activities. Thus, you can use this interface to track your models from inception to deployment.

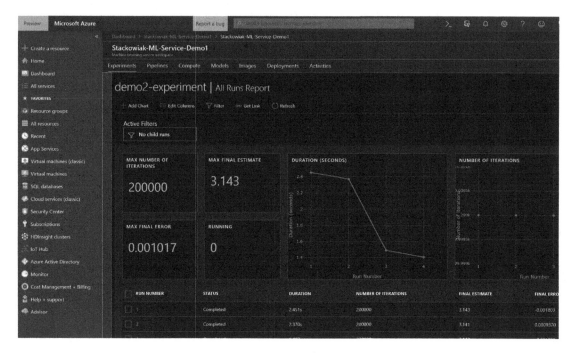

Figure 5-13. *Azure ML service experiment results tracked in the Azure Portal*

Cognitive Services

Azure Cognitive Services provides APIs, SDKs, and services enabling software developers to add cognitive features into applications. As noted in Chapter 2, these services focus in the areas of vision, speech, language, search, and decision. In the building of IoT applications, vision and decision are most often considered for deployment.

The Computer Vision Service provides advanced algorithms for processing information and returning information. The Custom Vision Service enables building of custom image classifiers. Both services are typically used with smart cameras that capture images at the edge and perform local analysis or transmit images to the cloud where the algorithms process the data.

The Computer Vision Service has several visual features relevant in IoT applications. It can be used to detect brands, assign images to categories based on taxonomies that you define, determine accent and dominant colors, provide descriptions, detect objects, and apply tagging.

The Custom Vision Service provides an image training environment. You begin by tagging a set of training images using tags that are consistent with what you are trying to detect. For example, if you are trying to train the service to detect the types of crops in a farm field, you'd first assemble a training set of images that are tagged with the crop types you wish to detect.

The image dataset in our example comes from public domain images posted by the USDA Agricultural Research Service (ARS). We tagged the images as showing alfalfa, corn, soybeans, or wheat. Figure 5-14 displays some of the images we uploaded into Custom Vision and denotes the number of each tagged image type used in the training.

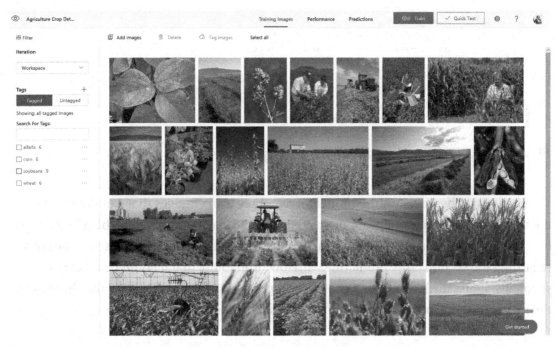

Figure 5-14. *Training images uploaded into Custom Vision*

Next you train the model and set a probability threshold for accuracy. The default is a goal of reaching 50 percent accuracy or above. You begin the training by simply hitting the train button shown in the previous figure.

Figure 5-15 illustrates the outcome of our second iteration of training. The precision indicates likeliness that a tag predicted by the model will be correct (60 percent in this example). Recall is a measurement of model sensitivity indicating the percentage of relevant tags detected (in comparison to the total relevant tags) and is also 60 percent in this iteration. The AP is average precision and is a measure of the model's performance summarizing the precision and the recall at different thresholds.

105

Figure 5-15. *Custom Vision image training performance*

Finally, you begin to test the model for accuracy using images that were not part of the training set. In the example shown in Figure 5-16, we have an image that has a couple of crops present. Corn is predicted with a high probability. Our model has less certainty regarding the second crop, predicting with low probability that it could be alfalfa or soybeans. If dissatisfied with this analysis, properly tag and add this and other images to the mix of training images and retrain the model producing a new iteration.

Image Detail   ✕

My Tags

| Add a tag and press enter |

Predictions

Tag	Probability
corn	93.4%
alfalfa	4.1%
soybeans	2.4%
wheat	0%

Save and close

Figure 5-16. *Custom Vision tested with an image not used in training*

Custom Vision can have many other use cases. For example, models might be produced for use in visual inspection of the condition utility lines to determine the need for their replacement, analyzing medical images for possible anomalies where further diagnoses might be needed, and determining whether there is proper alignment of components being placed into parts on a manufacturing assembly line.

Among the decision APIs, the Anomaly Detector is particularly relevant to IoT applications. You can use these RESTful APIs to detect anomalies in streaming data, leveraging previously seen data points. The APIs can also generate models that detect anomalies in JSON formatted time series datasets created in batch processes.

The APIs can provide details about the data including expected values, anomaly boundaries, and positions. Anomaly boundaries are automatically set. However, you can manually adjust the boundaries if you prefer more (or less) sensitivity in identifying anomalies.

Data Visualization and Power BI

Power BI is a business intelligence platform from Microsoft used in visualizing, aggregating, analyzing, and sharing data and data analysis. The Power BI service is deployed in the Microsoft cloud. The Power BI Desktop is free, downloadable software for your personal computer providing an environment to connect to data sources, develop data models, create visuals, and combine visuals into reports. Once created, you can publish these reports to the Power BI service.

When starting in Power BI Desktop, you likely will first download a sample of data to begin development. As development progresses and/or you deploy to the Power BI service, you can use Direct Query to analyze and report on the full live dataset.

In IoT scenarios, typical data sources include Blob Storage, Azure Data Lake Storage, HDInsight (HDFS, Interactive Query, and Spark), and Cosmos DB. Relational database sources that can be accessed include Azure SQL Database, Azure SQL Data Warehouse, Azure Analysis Service, Microsoft SQL Server and SQL Server Analysis Services, IBM DB2, Informix, and Netezza, MySQL, Oracle, PostgresSQL, SAP HANA and Business Warehouse, Snowflake, and any database supporting ODBC. Online services such as Dynamics and Salesforce can be accessed. Additionally, file types such as Excel, XML, JSON, PDF, and text or CSV can be leveraged.

Once loaded into Power BI Desktop, you might choose to transform data in the data model. For example, you can rename tables, update data types, append tables together and cleanse data so that similar sets can be combined, and rename groups of data.

You can model data relationships within the Power BI Desktop or rely on the Desktop to automatically infer relationships. Figure 5-17 illustrates the relationships that might exist in data coming from a smart retail shelf application that gathers information on products being put into shopping carts and identifies out of stock situations.

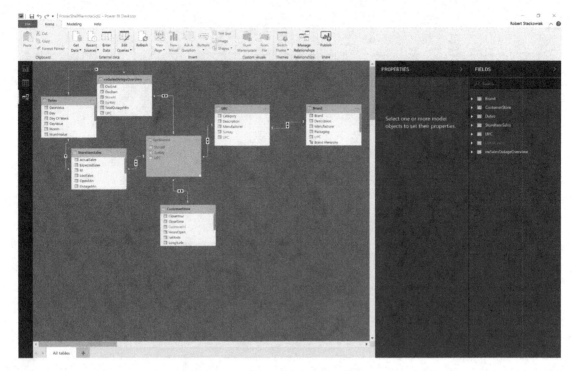

Figure 5-17. *Layout of tables in Power BI Desktop*

As you create the report, you can select from many different data visualizations provided. Examples of available visualizations include stacked bar charts, stacked column charts, clustered bar charts, clustered column charts, 100 percent stacked bar charts, 100 percent stacked column charts, line charts, area charts, stacked area charts, line and stacked column charts, line and clustered column charts, ribbon charts, waterfall charts, scatter charts pie charts, donut charts, treemaps, filled maps, funnels, gauges, cards, multi-row cards, KPIs, slicers, tables, matrices, R script visuals, Python visuals, ArcGIS Maps, globe maps, tornado charts, and custom visuals that you import.

A typical report created in the Power BI Desktop appears in Figure 5-18. We see a couple of visualizations from reported data on the left (table and line chart views), additional visualizations available and filters applied in the right center, and data items selected from the tables used in the report on the right.

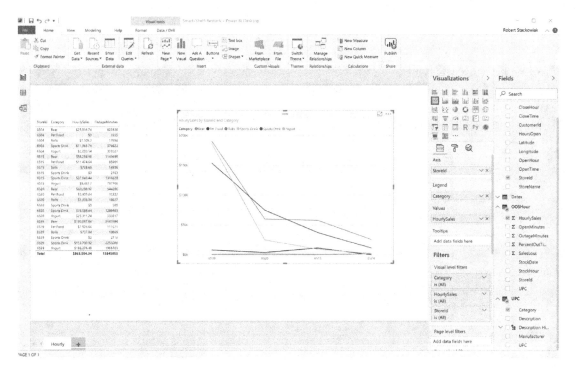

Figure 5-18. *A typical report in Power BI Desktop*

Reports are published to the Power BI service to enable access by a community of users. In Figure 5-19, we show what the same desktop report would initially look like in the Power BI service.

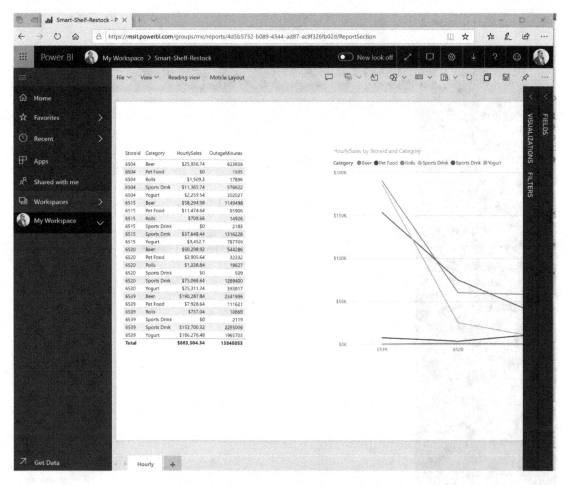

Figure 5-19. *A report rendered in Power BI in a web browser view*

Within the Power BI service, you can create a layout of the same report as it would appear on a mobile device as illustrated in Figure 5-20.

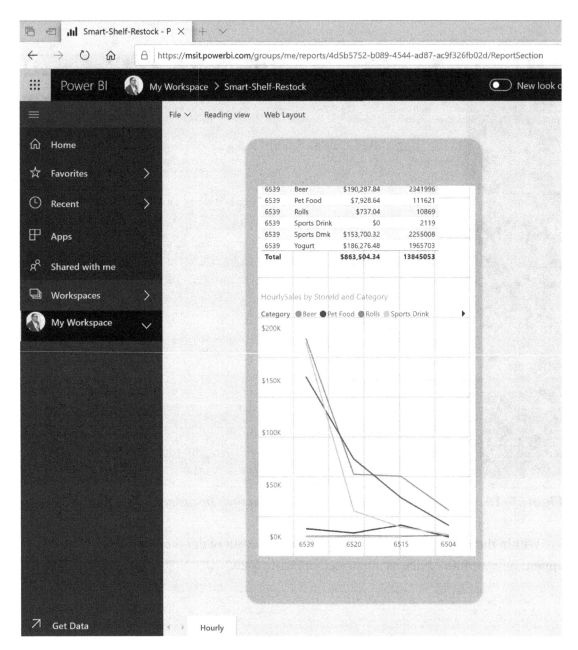

Figure 5-20. *A report rendered in Power BI as a mobile view*

From within the Power BI service, you can also create different reports applying other filters and visualizations. In Figure 5-21, we see creating a report focused on out-of-stock items and their impact on revenue.

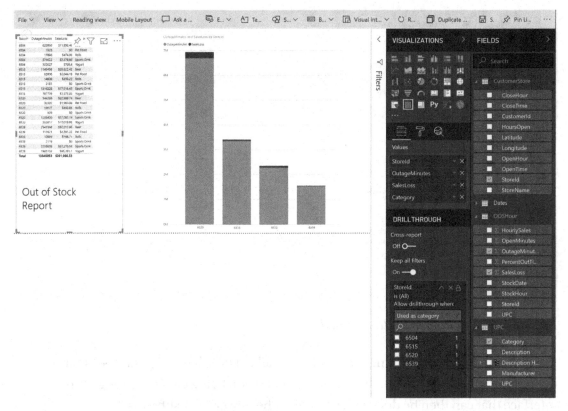

Figure 5-21. *A new report created in Power BI*

Whereas reports show data from a single dataset, dashboards can display data present from a variety of datasets and reports. As such, they can provide a more holistic view as to how a business is functioning and leverage data from IoT devices and lines of business systems.

Dashboards are created only in the Power BI service (not through the Desktop). The dashboards can be created from scratch directly from datasets, by pinning reports, or by modifying existing dashboards.

A supplier quality analysis Power BI dashboard appears in Figure 5-22 as an example. The dashboard presents data in tiles with a variety of visualizations present in this example.

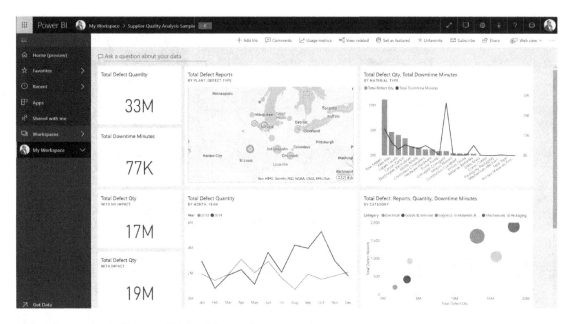

Figure 5-22. *A Power BI dashboard*

Power BI has a natural language interface called Q&A that can guide users through data exploration. Figure 5-23 illustrates a visualization being created through this interface that can then be deployed as a tile to the Power BI dashboard.

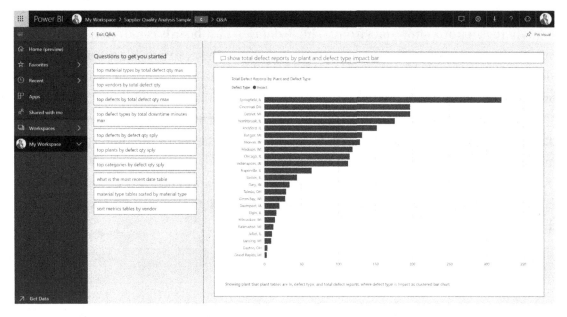

Figure 5-23. *Power BI dashboard Q&A*

Quick Insights can guide you toward interesting information in your data. You can run Quick Insights against datasets or individual dashboard tiles. The algorithms that are applied discover

- Category outliers (top and/or bottom)

- Change points in a time series

- Correlation

- Low variance

- Major factors (e.g., most of a total value comes from a single factor)

- Overall trends in time series

- Seasonality in time series

- Steady share

- Time series outliers

Sample output from Quick Insights against the data in our earlier smart shelf example produced various charts. Figure 5-24 shows average of outage minutes vs. hourly sales (with an outlier indicated) and count of manufacturers vs. hourly sales.

Quick Insights for **Smart-Shelf-Restock**

A subset of your data was analyzed and the following insights were found. **Learn more**

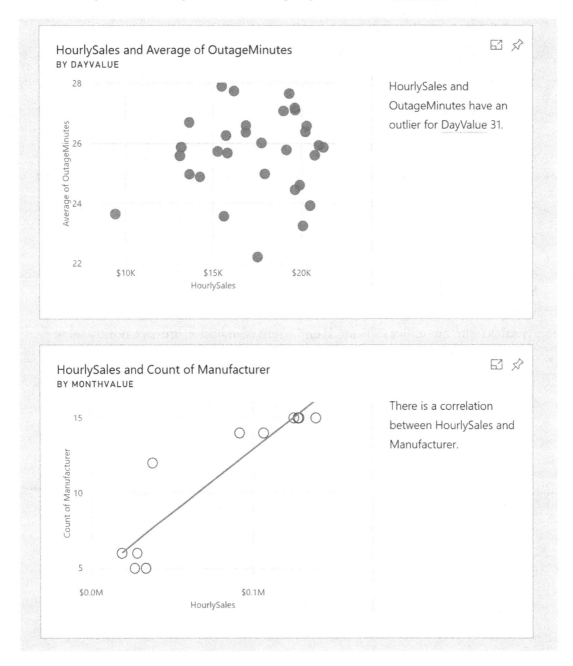

Figure 5-24. *Quick Insights output*

Note   Power BI users can be granted access to Azure Machine Learning models developed by data scientists. Power Query will discover the models which the user has access to and exposes them as dynamic Power Query functions. At the time this book was published, this capability was supported in Power BI dataflows and in Power Query online in the Power BI service.

You can collaborate with others in the creation of reports and dashboards by sharing workspaces. Once created, access to reports and dashboard tiles can be made available through Microsoft Teams by adding Power BI Tabs to channels and pointing to the report or tile link. Reports can also be printed (including as PDFs) or embedded into portals.

Reports and dashboards in the Power BI service can also be shared directly to e-mail addresses where the individuals will have the same access as the publisher (unless row-level security applied to the dataset restricts them). When granting access, the publisher can choose to allow the recipient to also share the report or dashboard or build new content using the underlying dataset.

Azure Bot Service and Bot Framework

Bots provide a question and answer or natural language interface akin to talking to a human or intelligent robot. The Azure Bot Service and Bot Framework provide tools used in building, testing, deploying, and managing intelligent bots. Microsoft provides an extensible framework that includes the SDK, tools, templates, and AI services.

You can extend your bot's functionality by using Microsoft's QnA Maker to set up a knowledge base to answer questions. Natural language understanding is accomplished by leveraging LUIS in Cognitive Services. Multiple models can be managed and leveraged during a bot conversation. Graphics, menus, cards, and buttons can be added to text to complete the experience.

For example, you might use QnA maker as a front-end to users that then pushes SQL to backend data management systems. You might also use a bot to push a command to an IoT edge device.

Figure 5-25 illustrates a quick start for setting QnA Maker up available through the Azure Portal.

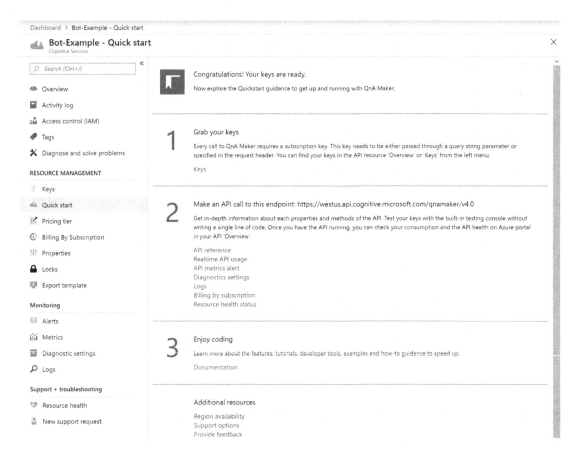

Figure 5-25. *QnA Maker quick start accessed through Azure Portal*

Microsoft provides a Bot Framework Emulator useful in debugging and interrogations. Once you have configured your bot in the Azure Portal, the bot can also be reached through a web chat interface for testing. When testing is complete, you can publish your bot to Azure or a web service.

Once deployed, you can gather data in the Azure Portal related to traffic, latency, users, messages, and channels. You can use this data to determine how best to improve the capabilities and performance of your bot.

CHAPTER 6

IoT Central and Solution Accelerators

Now that we've explored many of the key components in an IoT architecture, we are going to look at Microsoft's IoT solutions that package several of these components together. Each is designed to simplify and speed deployment of commonly implemented IoT solutions.

All the solutions described in this chapter feature the Azure IoT Hub. As discussed in Chapter 4, the IoT Hub enables connectivity and management that can be scaled to interface with large numbers of devices and enables high-volume telemetry ingestion, command and control of the devices, and enforcement of device security.

We'll begin by describing Microsoft's SaaS IoT offering called IoT Central. It is a Microsoft-managed offering in which underlying services are not exposed. Setup and management of IoT Central are via a browser-based interface.

The Microsoft IoT solution accelerators access a variety of underlying PaaS services and are designed to enable a greater degree of customization. When this book was published, the following solution accelerators were available:

- Remote monitoring

- Connected factory

- Predictive maintenance

- Device simulation

IoT Central and the solution accelerators are accessible via Microsoft web sites where you will also find links to documentation, developer's guides, an IoT School, the IoT Show (pre-recorded interviews/overviews describing component capabilities), access to the IoT Technical Community, and access to the IoT Device Catalog.

© Robert Stackowiak 2019
R. Stackowiak, *Azure Internet of Things Revealed*, https://doi.org/10.1007/978-1-4842-5470-7_6

As you might expect, the major sections of this chapter are the following:

- Azure IoT Central

- IoT Solution Accelerators

Azure IoT Central

Azure IoT Central provides a SaaS solution for gathering time series data from devices linked to the Azure IoT Hub and providing an interface for monitoring and managing the devices through Time Series Insights. It is designed to align with the roles and activities of the following individuals involved in your project:

- **Builders.** Define the types of devices connecting to the IoT Central application and customize the application. Builders create device templates to define telemetry that is being sent, define business and device properties, set thresholds that the application responds to, set device behavioral settings, and test the templates (often by initially using simulated data).

- **Operators.** Manage devices connected to the application including device monitoring, troubleshooting and remediation of problems, and provisioning of new devices.

- **Administrators.** Manage access to the IoT Central application through user roles and permissions.

- **Device Developers.** Create code to run on the devices using SDKs. The code is used in creating secure connections, sending telemetry, reporting on status, and receiving configuration updates.

IoT Central is accessible through a Microsoft web site at `https://azure.microsoft.com/services/iot-central`. The web site heading is pictured in Figure 6-1. This site provides the links and information you need to get started.

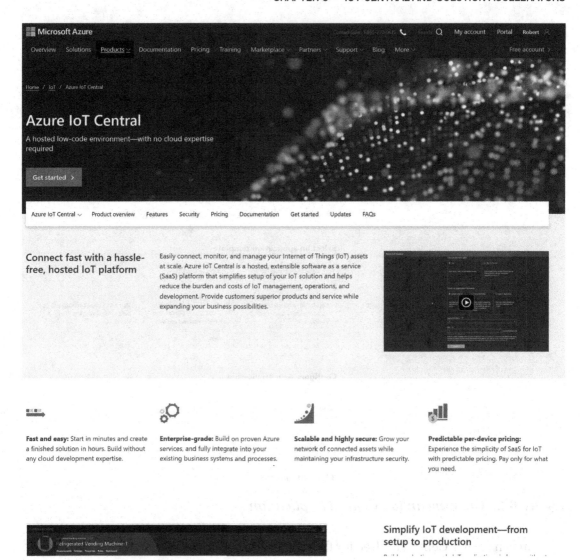

Figure 6-1. *IoT Central web site*

When you are ready to build your first application and initially enter IoT Central, you'll be presented with the screen shown in Figure 6-2.

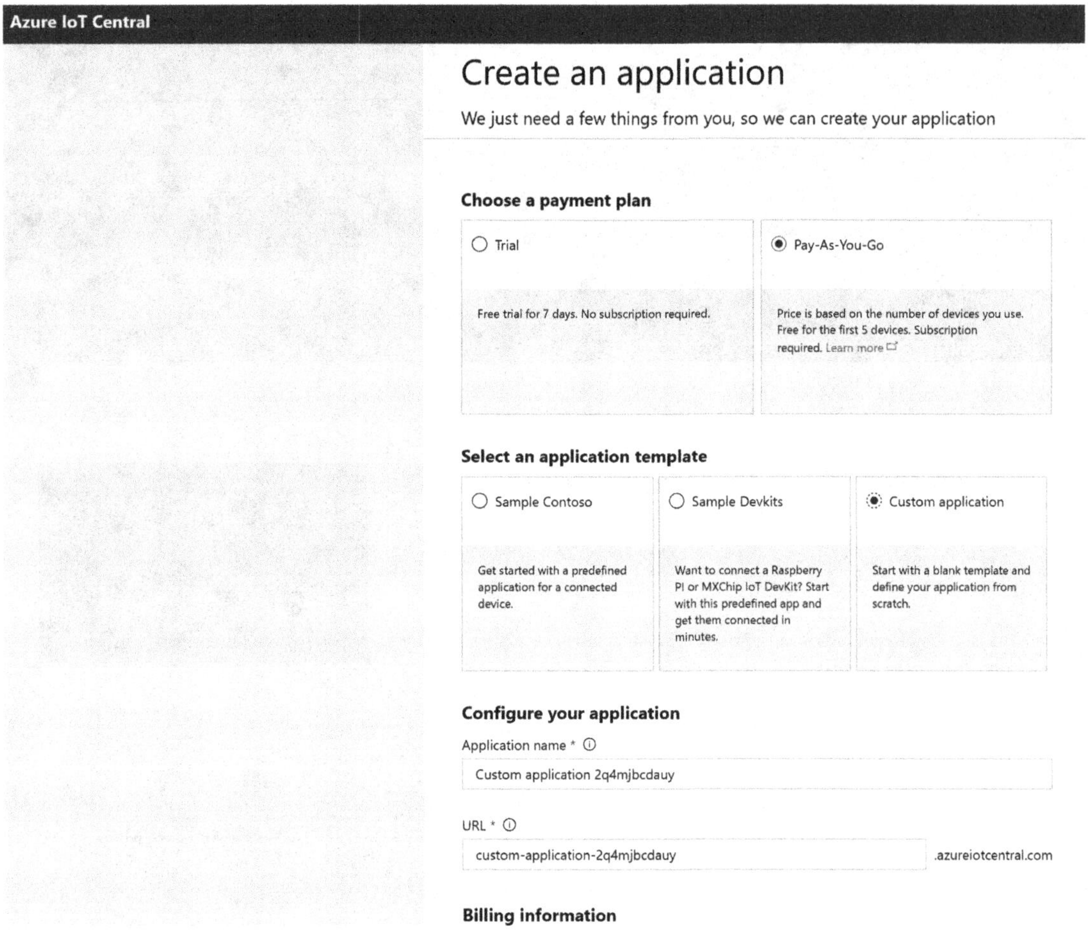

Figure 6-2. *Creating an IoT Central application*

You can try IoT Central for free for the first 7 days or choose to pay as you go. Then you select an application template from samples that are provided or define your own custom application. If new to IoT Central, you might choose to deploy the Contoso sample template that Microsoft provides so that you can gain familiarity with the subsequent interfaces.

We've selected creating a custom application in the previous figure. We also gave the application a name, noted the URL assigned, and provided appropriate billing information. The following dashboard and options are then presented as shown in Figure 6-3.

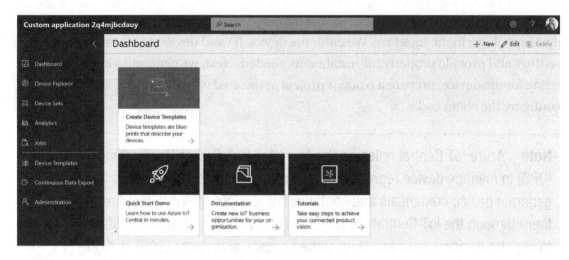

Figure 6-3. *Creating a custom IoT Central application*

Choosing to create device templates, we next define our device measurements (telemetry, state, event, and location), settings for devices (numbers, text, date, toggle, section labels), properties (such as device, customer, and service information), commands that remotely manage devices, rules that trigger actions when certain monitored conditions arise, and the dashboards for our devices. The interface to do this in IoT Central, including the measurements that can be created, is shown in Figure 6-4.

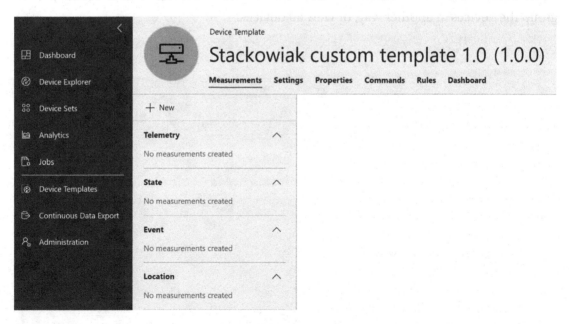

Figure 6-4. *Interface to create a device template for an IoT Central application*

Once a new device type is defined in a template, we can add the device using the device explorer in the interface. We enter the device ID and device name and then adjust settings and provide property information as needed. Next, we generate a connection string for the device, prepare a Node.js project associated with the device, and then configure the client code.

Note   Azure IoT Central relies on the Azure IoT Hub Device Provisioning Service (DPS) to manage device registrations and connections to your devices. You can generate device credentials and configure the devices offline without registering them through the IoT Central interface and use your own device identifiers to register devices. You can set up shared access signatures (SAS) or X.509 certificate authority to enable devices to connect. All data exchanged between the devices and Azure IoT Central is encrypted.

For purposes of illustrating what a deployed solution looks like, we've created the Contoso sample application that gathers data from simulated refrigerated vending machines. If you choose to create that sample application, your dashboard should appear like the image shown in Figure 6-5. There are options to view just devices with active maintenance contracts, just the devices located in Seattle, add a device set to group the devices in another way, or view all devices.

Figure 6-5. *IoT Central application dashboard for Contoso sample application*

On the menu along the left side, in addition to viewing the dashboard, we can access device explorer, device sets, analytics, and jobs. We also see access to device templates, continuous data export, and administration.

In the sample application, if we access device explorer and look at the measurements for one of the devices, we'll see telemetry measurements being tracked on the left (including accelerometers, gyroscopes, humidity, magnetometer, pressure, and temperature). We also see the time series display of recent measurements tracked in the chart on the right as shown in Figure 6-6.

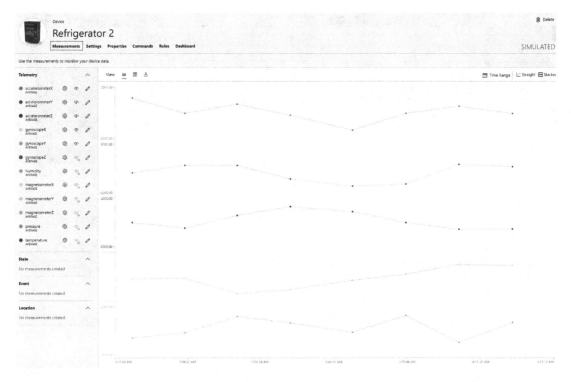

Figure 6-6. *Sample measurements shown in IoT Central Device Explorer*

The dashboard in device explorer consists of customizable views of a device. For the same refrigerated vending machine in our example, the dashboard is set up to show machine information; anti-tampering information based on data from the accelerometers; maximum temperature, average pressure, and minimum humidity readings; and a chart of environmental trending over time of humidity, temperature, and pressure. That dashboard is shown in Figure 6-7.

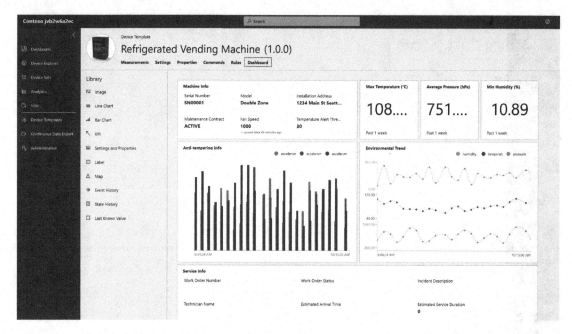

Figure 6-7. *Sample refrigerator dashboard in IoT Central Device Explorer*

In device sets, we find the device set names, their descriptions, and the device template used. We can set conditions for device sets by selecting properties (such as location, temperature alerts, fan speed, etc.), an operator (value for the property equals, does not equal, is greater than, is greater than or equal to, is less than, is less than, is equal to, contains, etc.), and a value. We can also view individual devices within each device set. Here, we can view measurements, settings, properties, commands, rules, and the dashboard for each device. For example, we might set a rule that an alert is sent if the device is moved more than a certain distance.

In analytics, we choose the device set we want to apply analytics to; set a filter on a measurement including a condition, operator, and value (similar to device sets approach); define a time period we want to look at (from 10 minutes to last month to custom); and then show results. Figure 6-8 shows a view of analytics performed based on filtering applied to temperature readings.

Figure 6-8. *IoT Central analytics applied to refrigerator temperature data*

Jobs can be created to set device properties or settings. We begin by defining a job name and providing a description and device set to use. We define whether the job type is aligned to device properties or settings and then select a property or setting and provide a value. Figure 6-9 shows setting up a job to set fan speed in a couple of the refrigerated vending machines in a defined group (i.e., those that are in Seattle).

Figure 6-9. *Interface to create a job in an IoT Central application*

Functions below the line on the left side of the application interface in the previous figures are focused on setup and management. These include access to setting up device templates (previously introduced), continuous data export, and administration.

When creating a new device template, you can create a custom template or utilize one of those that Microsoft provides as shown in Figure 6-10. A new device template is created as version 1.0.0.

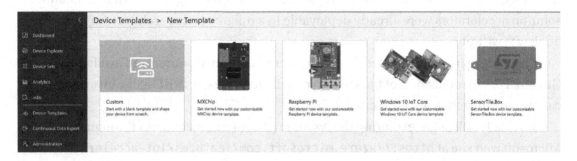

Figure 6-10. *Creating a new device template in IoT Central*

If you later make changes to settings or required properties in the device template, you will be prompted to create a new version. This can be extremely useful as you might initially find that rules are broken in the new version, such as when properties that conditions rely upon have been removed. Some of the tiles in your dashboard might also be broken if properties or settings are removed. While you fix these problems, operators will still have access to the old fully working version. When you are ready, you can migrate devices to the new version through device explorer.

Continuous data export enables you to export data from IoT Central to your storage. (i.e., Azure Blob Storage, Azure Event Hubs, Azure Service Bus). The administration interface enables management of application settings, users, roles, billing, device connection, access tokens, application customization, help customization, and application template export.

IoT Solution Accelerators

The IoT solution accelerators are designed to speed implementation of popular IoT scenarios, such as those for remote monitoring, connected factories, and predictive maintenance, by automatically provisioning key PaaS Azure cloud services needed in each scenario. Microsoft positions the solution accelerators as starting points for your own IoT solutions. They are designed to be scalable, modular, understandable, extensible, and secure.

At the time this book was published, Microsoft was in the process of moving the solution accelerators from a model-view-controller (MVC) architecture written in .NET to a microservices architecture. A microservices architecture can improve the flexibility, reliability, and scalability of a solution. The Remote Monitoring and Device Simulation solution accelerators were already deployable in a microservices architecture when this book was written.

The underlying code in the solution accelerators is open source and available on GitHub. For customization of backends, you will need Java or .NET skills. Visualizations can be customized using JavaScript.

You likely will begin exploring the Azure IoT solution accelerators through the Microsoft web site at `https://azure.microsoft.com/features/iot-accelerators`. Figure 6-11 shows the web site heading.

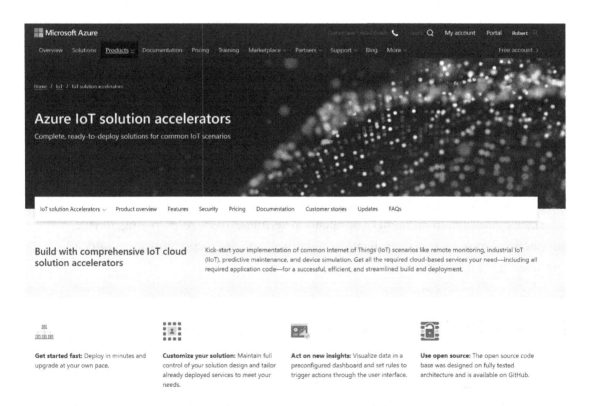

Figure 6-11. *IoT solution accelerators home page*

Figure 6-12 shows the four solution accelerators on the home page. By choosing links in this interface, you can provision each of the services. If you prefer, you can also deploy the solution accelerators from the command line. For solution accelerators other than Device Simulation, there is also a link to try a demo of the solution. The demos can be particularly useful when you want to explore the capabilities that these solution accelerators provide.

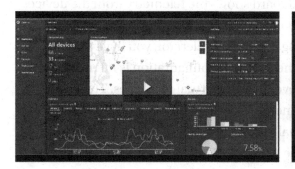

Azure IoT for remote monitoring

Collect and analyze real-time device data to trigger automatic alerts and actions—including performing remote diagnostics and automatically initiating maintenance requests. Use the remote monitoring dashboard to view telemetry from your connected devices, provision new devices, and upgrade firmware on your connected devices.

Learn more >

Try the demo >

Provision the service >

Resources and services used   ∨

Azure IoT for device simulation

Develop and test your IoT solution on simulated devices using the device simulation solution accelerator. Conduct realistic tests using complex device models to emulate real-world scenarios. Use the device simulation web app to configure and run simulations.

Provision the service >

Resources and services used   ∨

Azure IoT for connected factory

Connect and monitor your industrial assets using standards like OPC-UA with the Azure IoT connected factory solution accelerator. Extract data from brownfield devices to start gathering insights to drive increased performance on the factory floor. Monitor and manage your industrial devices using the connected factory dashboard.

Learn more >

Try the demo >

Provision the service >

Azure IoT for predictive maintenance

Analyze streaming data from sensors and devices to predict equipment failures and avoid costly repairs. Go beyond monitoring assets and catch potential issues before they turn into problems. View predictive maintenance analytics using the dashboard.

Learn more >

Try the demo >

Provision the service >

Figure 6-12. *IoT solution accelerators*

Next, we'll explore each of the IoT solution accelerators regarding their capabilities and key resources that are provisioned.

Remote Monitoring

The Remote Monitoring solution accelerator enables collection of telemetry from multiple devices in remote locations. A dashboard shows the telemetry from the devices and provides an interface used to provision new devices or upgrade device firmware.

When you deploy the Remote Monitoring solution accelerator, you have a choice of standard, basic, or local configurations. The standard configuration is intended for production and deploys microservices on several Azure virtual machines. The basic configuration is intended for testing and demos and deploys the microservices on a single Azure virtual machine. The local virtual machine deployment is intended for testing and development and connects to the Azure IoT Hub to reach cloud resources.

Automated provisioning provided by the Remote Monitoring Solution Accelerator generates, creates, and configures the activities used in setting up the needed cloud services. These activities appear in Figure 6-13. In our example, we built the basic configuration.

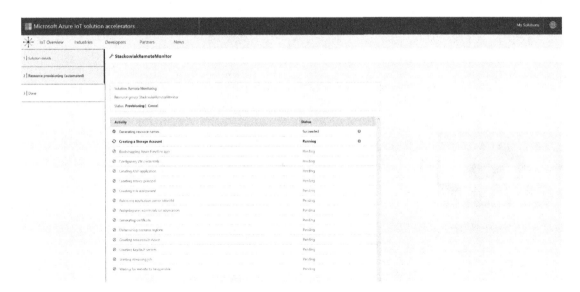

Figure 6-13. *Remote Monitoring solution accelerator automated provisioning*

The Azure cloud services started in the resource group that are created are shown in Figure 6-14. Among the key resources made available is the single virtual machine for the microservices, the Azure IoT Hub, Cosmos DB, storage accounts, Time Series Insights, Azure Maps, Stream Analytics job, Event Hub, Logic App, and Apps Services. Microservices in this solution include an IoT Hub Manager microservice, device telemetry microservice, storage adapter microservice, Azure Stream Analytics manager microservice, and device simulation microservice.

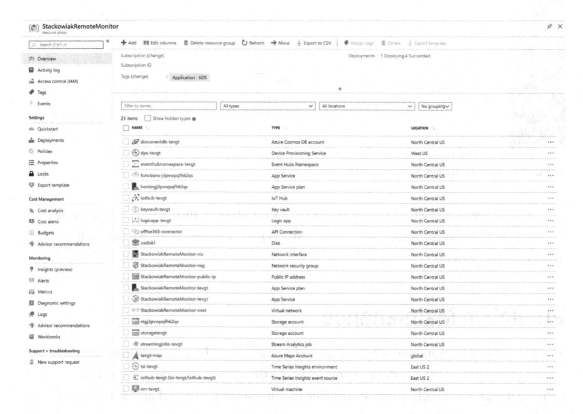

Figure 6-14. *Azure Resource Group for Remote Monitoring services*

When using the Remote Monitoring solution accelerator for demos (along with the supplied simulated device data), you can explore readings from a variety of chillers, elevators, engines, trucks, and prototypes. The main dashboard is shown in Figure 6-15.

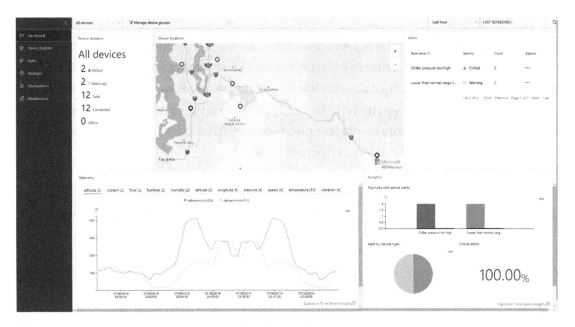

Figure 6-15. *Remote Monitoring dashboard*

You can see that the dashboard leverages Azure Maps to display where the devices are located. Integration with Time Series Insights (e.g., the menu shown on the left in the figure) is evident.

Predictive Maintenance

The Predictive Maintenance solution accelerator uses machine learning algorithms applied to device telemetry data to predict when the devices will fail. This solution can be used to put into practice optimal device maintenance plans and activities.

Automated provisioning provided by the Predictive Maintenance solution accelerator generates, creates, and configures the activities used in setting up the needed cloud services. These activities appear in Figure 6-16.

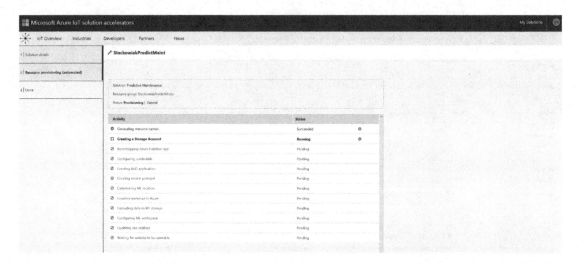

Figure 6-16. *Predictive Maintenance Solution Accelerator automated provisioning*

The Azure cloud services started in the resource group that are created are shown in Figure 6-17. Key resources made available include the Azure IoT Hub, storage accounts, a Machine Learning Studio workspace, Stream Analytics job, Event Hub, and Apps services. The Stream Analytics job first selects all device telemetry and sends data to blob storage for visualization and then computes average sensor values over 2-minute sliding windows (sending this data through an Event Hub to an event processor).

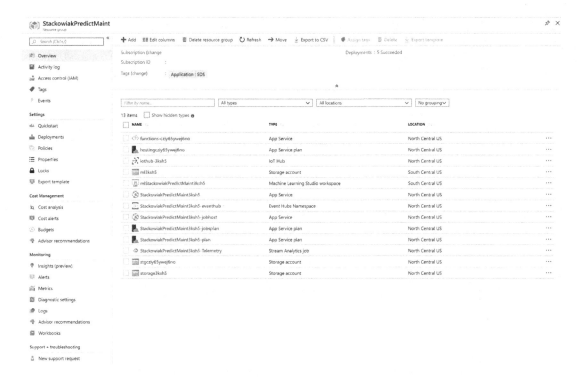

Figure 6-17. *Azure Resource Group for Predictive Maintenance services*

Once provisioned, a link is provided to the Machine Learning Studio and the workspace.

Figure 6-18 shows a view of a model provided for demonstration purposes that includes a regression algorithm developed by Microsoft using a public sample data set that contains telemetry coming from sensors in jet engines.

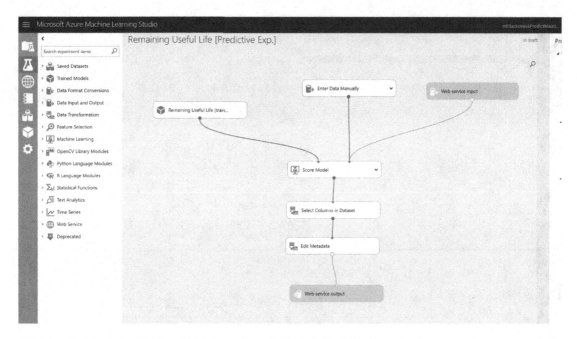

Figure 6-18. *Predictive Maintenance ML Studio Workspace*

The Predictive Maintenance solution accelerator also includes a demo dashboard for the same sample data set. The regression algorithm that is deployed predicts the Remaining Useful Life (RUL) of the two jet engines as data from four sensors in each engine is cycled through. Each cycle denoted in the dashboard represents a flight of 2 to 10 hours. Data is captured by sensors every 30 minutes during a flight.

Figure 6-19 shows the RUL dashboard displaying these KPIs including charts of recent readings and predictions.

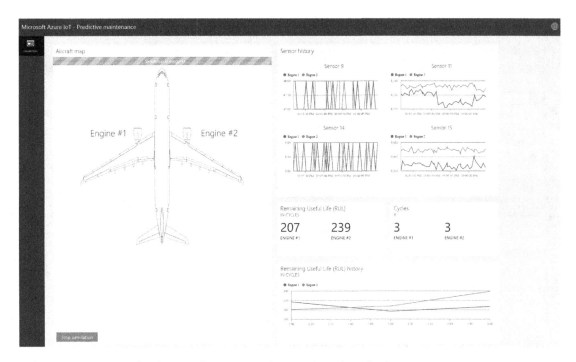

Figure 6-19. *Predictive Maintenance jet engine simulation*

Connected Factory

The Connected Factory solution accelerator enables you to spin up in an automated fashion the resources needed in deploying an Industrial Internet of Things footprint. The industrial devices connect through the OPC UA interface. A cloud dashboard is part of the implementation and provides the following functionality:

- Enables browsing of the OPC UA information model in OPC UA servers

- Enables configuration of OPC UA devices (call methods, read and write data)

- Enables publishing/unpublishing OPC UA device telemetry data

- Enables viewing of telemetry previews

- Enables viewing of telemetry data trends and creation of correlations using Time Series Insights

- Enables viewing of calculated overall equipment efficiency (OEE) and key performance indicators

- Enables viewing of industry asset hierarchies in tree topologies and interactive maps

- Enables viewing, acknowledgment, and closing of alerts based on threshold rules that you set

Security permissions for users are configured based on role-based access control (RBAC). End-to-end encryption is implemented using OPC UA authentication (X.509 certificates) and security tokens.

Automated provisioning provided by the Connect Factory solution accelerator generates, creates, and configures the activities used in setting up the needed cloud services. These activities appear in Figure 6-20.

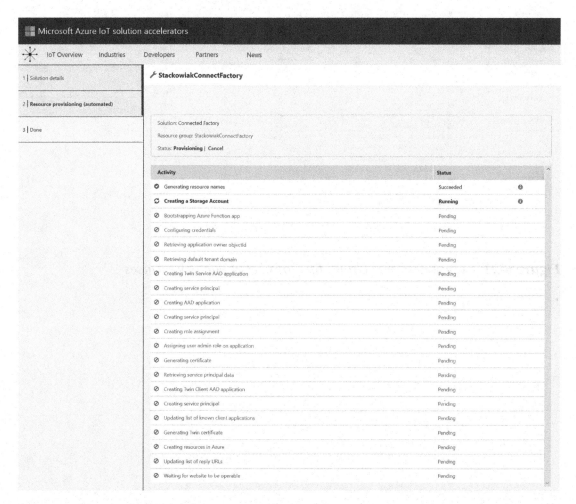

Figure 6-20. *Connected Factory Solution Accelerator automated provisioning*

The Azure cloud services started in the resource group that are created are shown in Figure 6-21. Key resources made available include the Azure IoT Hub, Cosmos DB, storage accounts, Time Series Insights, Azure Maps, Event Hub, and Apps Services.

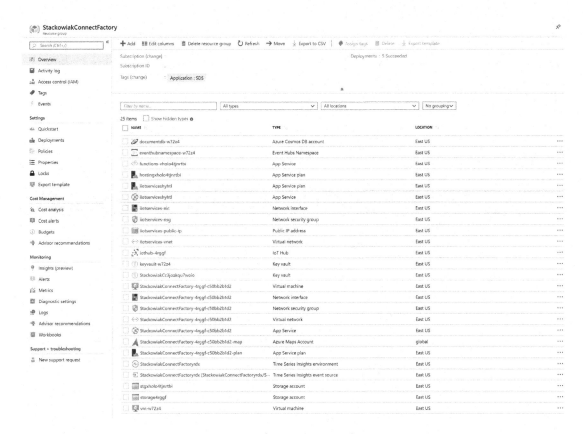

Figure 6-21. *Azure Resource Group for Connected Factory services*

A demonstration dashboard is provided with simulated device data that can help you better understand the functionality provided. A map provides a view of where factories are located. The status of each factory is shown as is a list of current alarms. Overall equipment efficiency, availability, performance, quality, units per hour, and kWh are represented by indicators. You can then drill to further detail through the dashboard.

Figure 6-22 shows a couple of the radial gauge charts presented in the dashboard.

Figure 6-22. *Connected Factory dashboard radial gauge charts*

Device Simulation

The Device Simulation solution accelerator is designed to define simulated devices that create realistic telemetry. The telemetry can then be used in testing IoT solutions that you are developing. The modeling includes message formats, twin properties, and methods. More complex device behaviors can be simulated using JavaScript.

You can simulate a single device during testing or scale the testing to thousands of devices connected to your IoT Hub(s). So, you can simulate normal, peak, and extreme workloads for scale testing.

Automated provisioning provided by the Device Simulation solution accelerator generates, creates, and configures the activities used in setting up the needed cloud services. These activities appear in Figure 6-23.

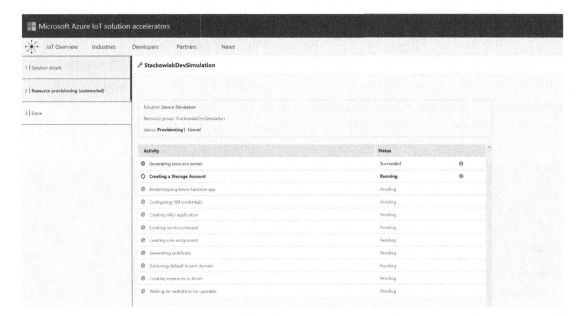

Figure 6-23. *Device Simulation Solution Accelerator automated provisioning*

The Azure cloud services started in the resource group that are created are shown in Figure 6-24. Key resources made available include the Azure IoT Hub, the Cosmos DB, the storage accounts, and the application services.

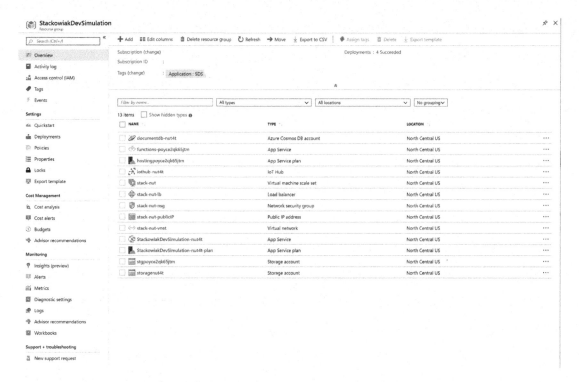

Figure 6-24. *Azure Resource Group for Device Simulation services*

Once provisioned, you can run the sample simulations provided (including good and faulty chillers, elevators, engines, trucks, and prototypes). You can also choose to define custom device simulations including data points and value ranges. As noted earlier, your third choice is to create advanced device simulations in JSON definition files using JavaScript that you can upload.

A sample dashboard is provided to view the simulations and provides information on the number of devices, the total messages and message rate, the number of failed messages, the number of device connections, and the number of failed twin updates.

CHAPTER 7

Infrastructure Integration

Thus far, our more detailed exploration of Microsoft's IoT footprint has focused on analysis of streaming data from devices, managing and monitoring the devices, and pushing intelligence to the edge. Chances are, you will want to integrate some of your existing data sources into your new IoT solution. Much of this chapter focuses on methods for finding and integrating the data that you will need.

We'll first look at typical preexisting potential sources of data. Then we'll explore the roles of Azure Data Factory, Azure Data Explorer, and PolyBase in moving and/or accessing data. We'll also look at configuring VPN connections and/or ExpressRoute for network connectivity between on-premises devices and systems and the cloud, using the Azure Data Box to physically move data to Azure data centers, and using Azure Data Catalog to find where data is located.

Finally, there are multiple Microsoft partners who provided data historians (that are time series databases) for years that are often part of IoT solutions. We'll look at how two of them, OSIsoft and PTC, are integrating Azure into their modern cloud-based deployment architectures.

The chapter includes the following major sections:

- Preexisting sources of data
- Integrating and finding data sources
- Data historians and integration with Azure

Preexisting Sources of Data

Organizations that deploy IoT solutions almost always find a need to integrate data from legacy transactional and data warehousing systems in order to deliver the key metrics needed in answering business questions. As transactional data fits neatly into rows and columns, relational databases are typically deployed to provide data management solutions for such data.

Preexisting footprints are almost always unique. Legacy data sources might reside on premises, in clouds, or from a combination of locations. On-premises relational databases typically found include Microsoft SQL Server, Oracle, IBM DB2, MySQL, PostgreSQL, and others. When deployed in Azure, these databases can be deployed within virtual machines. There are also databases available for PaaS deployment such as Microsoft SQL Database, Microsoft SQL Data Warehouse, and Snowflake.

Microsoft's Azure SQL Database is a fully managed service and is based on the latest version of Microsoft SQL Server general-purpose database engine. It is where you will find the newest capabilities for the SQL Server family released first. Azure SQL Database also supports non-relational structures including graphs, JSON, spatial, and XML. It has a hyperscale service tier that enables database scalability up to 100 terabytes. By configuring elastic pools, you can assign resources that are shared.

The Microsoft SQL Data Warehouse uses a massively parallel processing (MPP) engine to perform queries across extremely large databases often found in data warehouses, including those up to Petabytes in size. The database consists of a control node that optimizes and coordinates parallel queries and multiple compute nodes (up to 60).

In Figure 7-1, we illustrate a typical footprint that varies slightly from diagrams presented earlier in this book as we have identified some of the key transactional systems. In this figure, the ERP system might be SAP, Oracle E-Business Suite, Oracle Fusion Applications, Infor, or some other vendor solution that is deployed either on-premises or in a cloud. The ERP DW pictured would likely be SAP BW in an SAP implementation. The HR system could be Workday, another cloud-based HR solution, or a legacy HR system on premises. The CRM solution could be Microsoft Dynamics 365 or SalesForce in the cloud or other legacy on-premises or cloud application.

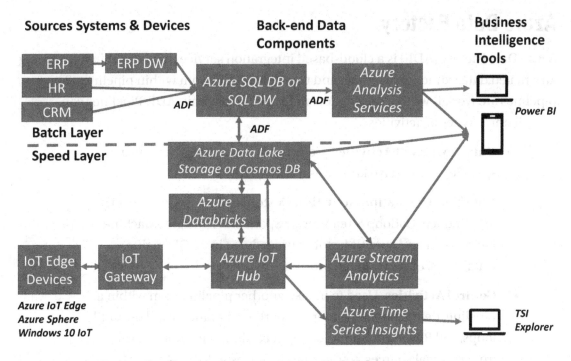

Figure 7-1. *Example of transactional systems in full footprint*

In this figure, we've pictured Azure SQL Database or Azure SQL Data Warehouse serving as data warehouses in the cloud. However, the data warehouse engine could also be one of the other cloud-based or on-premises databases that we previously mentioned.

Integrating and Finding Data Sources

We might choose to analyze our various sources of data by gathering the data to a single location. We could also keep the data in place and analyze it through a distributed query. In this section of the chapter, we explore ways of doing this and describe how you can track where the data resides. We begin by discussing how Azure Data Factory can be used to gather data.

Azure Data Factory

Azure Data Factory (ADF) is a cloud-based integration service that is used in performing automated data extraction, loading, and transform (ELT) from within pipelines. The pipelines are workflows that are created and scheduled within ADF. Pipelines can contain three types of activities:

- **Data Movement Activities.** Copy data from a data source to a specified target (also known as a data sink).

- **Data Transformation Activities.** Custom coding using Hive, Pig, MapReduce, Hadoop Streaming or Spark in HDInsight, Machine Learning in an Azure VM, stored procedures in a SQL engine, Databricks, or Azure batch process.

- **Control Activities.** Used to invoke another pipeline from within a pipeline, define repeating loops to perform iterations and do-until loops, call REST endpoints, lookup records, table names or values from external sources, retrieve metadata, establish branches based on conditions, and specify wait times for pipelines.

Data connectors are available for a variety of sources and targets in Azure including databases, NoSQL databases, files and file systems, and services and applications. You can also use generic protocols and interfaces to access data, such as ODBC, OData, and REST.

Figure 7-2 shows an Azure Portal interface view of linked Azure data management services that can serve as sources or targets. Azure data connectors shown include Azure Blob Storage, Cosmos DB, Azure Data Lake Storage, Azure Database for MariaDB, Azure Database for MySQL, Azure Database for PostgreSQL, Azure SQL Data Warehouse, and Azure SQL Database.

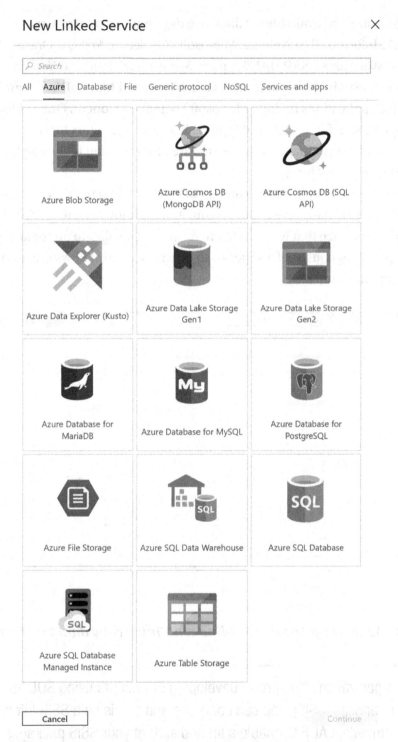

Figure 7-2. *Azure linked services within Azure Data Factory*

Connectors are also available for database data sources such as Greenplum, HBase, IBM DB2, IBM Informix, IBM Netezza, Microsoft SQL Server, MySQL, Oracle, PostgreSQL, SAP Business Warehouse, SAP HANA, Spark, Sybase, Teradata, and Vertica. Connectors for NoSQL databases include Cassandra, Couchbase, and MongoDB. Examples of connectors for applications include Microsoft Dynamics, Concur, Oracle Eloqua, Marketo, PayPal, Salesforce, SAP, and Square. Various connectors for the Amazon and Google cloud data management systems are also provided including for Amazon Redshift and Google BigQuery.

A simple Copy Activity function can be initiated in a data pipeline to move data from on-premises or cloud-based data sources into Azure for further processing. Figure 7-3 shows the ADF interface that we used to create and then validate our data copy pipeline. More often, pipelines consist of a series of steps where control flows are used to orchestrate pipeline activities.

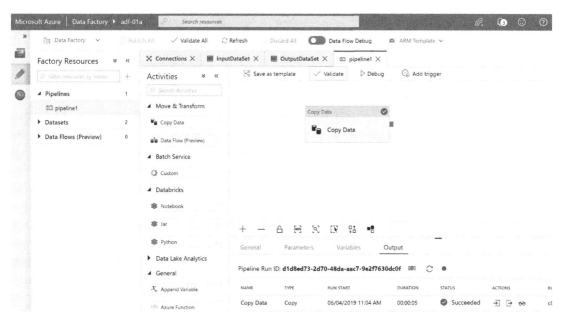

Figure 7-3. *Azure Data Factory pipeline containing copy data function*

Note   In organizations that prefer developing ELT scripts using SQL Server Integration Services (SSIS), you can configure and provision a SSIS Integration Runtime from within ADF to enable a lift and shift of your SSIS packages.

We can schedule the workflow for execution when we are satisfied with our design. Pipeline runs can be instantiated by passing arguments to parameters (e.g., datasets or linked services) using manual methods or within triggers.

Once deployed, you can monitor the success of your activities and pipelines. ADF supports pipeline monitoring via the Azure Monitor and APIs, PowerShell, Azure Monitor logs, and Azure Portal health indicators.

Query Services Across Diverse Data

Microsoft offers a couple of services enabling queries across semi-structured and structured data sources. Azure Data Explorer requires moving data to a common database, while PolyBase simply accesses data through external tables as if the data was local.

Microsoft describes Azure Data Explorer as a fast and scalable data exploration service. The cluster and database service can load streaming incoming data and/or copy data from other data sources through Event Hubs, Event Grids, Kafka, Python, Node SDKs, the .NET Standard SDK, Logstash, and ADF.

You begin by creating a cluster (including name and compute specifications) in a resource group using the Azure Portal. You then create a Data Explorer database and define a retention period (in days) and a cache period (in days). You start the cluster (and can later stop it to minimize cost). You next create a target table in the Azure Data Explorer cluster using a create table command through the query interface and provide a mapping of incoming JSON data to column names and data types in the table. Lastly, you add a data connection.

You can query data using Azure Data Explorer in a web-based interface. You can also write queries using the Kusto Query Language (KQL). Queries generally begin by referencing a table or function. Some of the advanced capabilities of KQL include scalar and tabular operators, time series analysis and anomaly detection, aggregations, forecasting, and machine learning.

PolyBase is built into SQL Server instances and the Azure Data Warehouse. You access data that resides in Azure Storage, Hadoop, and other file systems and databases using external table definitions. Connections to the data repositories are via ODBC. The definitions and user permissions are stored in the database. Query performance can be improved by configuring PolyBase scale-out groups.

Figure 7-4 illustrates a typical scale-out group consisting of a head node and multiple compute nodes. The illustration of the head node shows a SQL instance, PolyBase engine, and PolyBase Data Movement Service (DMS). The head node is the location to which queries are submitted. The PolyBase engine parses queries on external data, generates the query plan, and distributes work to the compute nodes' DMS. Compute nodes simply contain a SQL instance and PolyBase DMS and are replicated in numbers to adequately scale processing that will meet query performance needs.

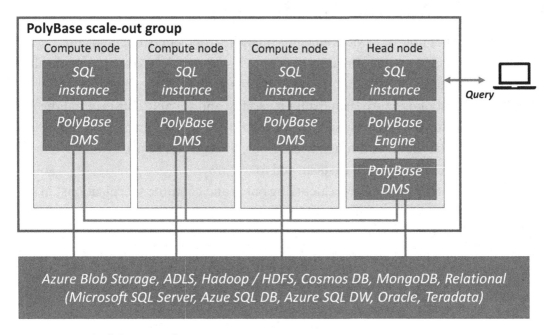

Figure 7-4. *PolyBase scale-out group*

PolyBase is sometimes used in combination with ADF, especially when the goal is to speed data transformations and updates. Behind the scenes, it provides the copy capabilities that we described earlier in the ADF section of this chapter.

Note   Key PolyBase features, including support for certain sources, vary based on the version of the SQL engine that PolyBase is bundled with. Always first check the Microsoft documentation regarding feature availability for the version that you plan to deploy.

Connecting On-Premises Networks to Azure

There are multiple ways by which you can connect on-premises networks to an Azure Virtual Network (VNet). They include

- Virtual Private Network (VPN) connection

- Azure ExpressRoute connection

- ExpressRoute connection with VPN failover

A VPN connection includes an Azure VPN gateway that is used to send encrypted traffic between the Azure VNet and an on-premises network. VPN connections use the public Internet and were limited to speeds of 1.25 Gb per second at the time this book was published. Network traffic received by the VPN gateway is routed to an internal load balancer when traffic is being sent to applications in Azure.

Figure 7-5 shows the Azure Portal interface used in configuring a VPN Gateway. Note that there is also an option to choose ExpressRoute as the gateway type which then presents a different set of prompts.

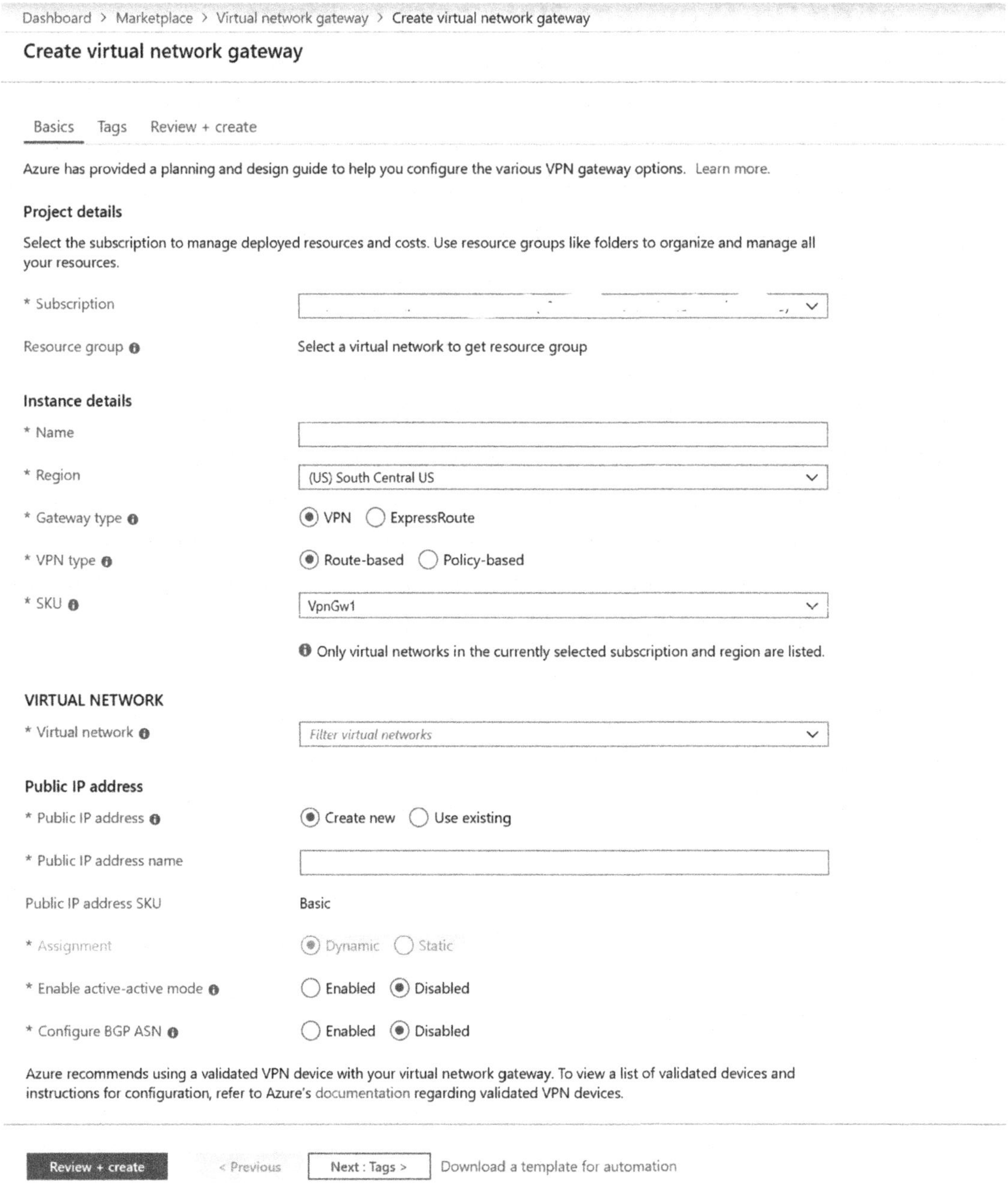

Dashboard > Marketplace > Virtual network gateway > Create virtual network gateway

Create virtual network gateway

Basics   Tags   Review + create

Azure has provided a planning and design guide to help you configure the various VPN gateway options. Learn more.

Project details

Select the subscription to manage deployed resources and costs. Use resource groups like folders to organize and manage all your resources.

* Subscription	[_____ ‑, ∨]
Resource group ❶	Select a virtual network to get resource group

Instance details

* Name	[_____]
* Region	[(US) South Central US ∨]
* Gateway type ❶	◉ VPN ◯ ExpressRoute
* VPN type ❶	◉ Route-based ◯ Policy-based
* SKU ❶	[VpnGw1 ∨]

❶ Only virtual networks in the currently selected subscription and region are listed.

VIRTUAL NETWORK

* Virtual network ❶	[*Filter virtual networks* ∨]

Public IP address

* Public IP address ❶	◉ Create new ◯ Use existing
* Public IP address name	[_____]
Public IP address SKU	Basic
* Assignment	◉ Dynamic ◯ Static
* Enable active-active mode ❶	◯ Enabled ◉ Disabled
* Configure BGP ASN ❶	◯ Enabled ◉ Disabled

Azure recommends using a validated VPN device with your virtual network gateway. To view a list of validated devices and instructions for configuration, refer to Azure's documentation regarding validated VPN devices.

[Review + create]   < Previous   [Next : Tags >]   Download a template for automation

Figure 7-5. *Configuring a VPN Gateway through the Azure Portal*

A VPN appliance must also be present on-premises providing external connectivity for that network. The VPN appliance might be a dedicated hardware device or a software service (such as the Routing and Remote Access Service in Windows Server).

Traffic over the public Internet is encrypted and flows through an IPSec tunnel. Given possible latency challenges when deploying VPN connections, they are generally used in situations where traffic is considered as being light. Figure 7-6 illustrates a VPN connection in a site-to-site configuration. Multiple VPN connections are typically present to help scale bandwidth.

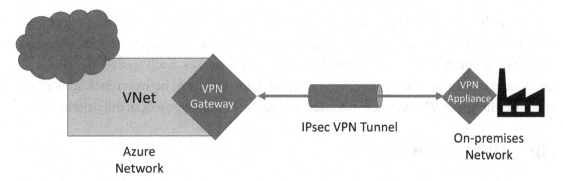

Figure 7-6. *VPN site-to-site configuration*

Azure ExpressRoute connections use private and dedicated two-layer or three-layer circuits provided by third-party network providers. Bandwidths of up to 10 Gb per second are possible. Some providers offer dynamic scaling of bandwidth to meet changing requirements and to enable charge-back for the bandwidth that is being used. High-bandwidth routers are required for connection to on-premises networks. Microsoft edge routers are used to provide connections to the Azure VNet in the cloud. Figure 7-7 illustrates private ExpressRoute connection in a site-to-site configuration.

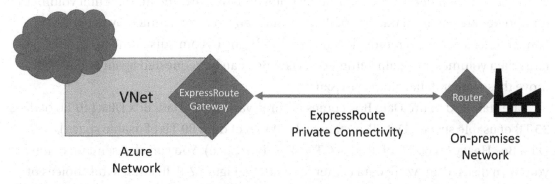

Figure 7-7. *ExpressRoute site-to-site configuration*

ExpressRoute with VPN failover can be thought of as combining the two previous options that we just described. Both types of connections are configured. ExpressRoute circuits are configured to provide connectivity under normal conditions. Failover to VPN connections are configured so that they can provide connectivity in situations where there is a loss of ExpressRoute connectivity.

Within the backend cloud, a hub–spoke topology can be deployed in Azure to isolate workloads when services are to be shared. For example, network virtual appliances and DNS servers can be shared for different workloads, different departments, or different stages of development and deployment.

The hub–spoke topology is deployed using CLI scripts found in GitHub. An Azure VNet serves as a hub in the topology providing a central connection to the on-premises network and provides shared services for other VNets serving as spokes. Each spoke VNet is connected to the hub VNet by peering, enabling traffic exchange between each spoke and the hub. The spoke VNets enable isolation and separate management of the workloads.

Bulk Data Transfer

Legacy data warehouses and remote IoT databases sometimes measure in the terabytes or more in size. Network connections into Azure might be too slow to provide timely and reliable data transfers. In such scenarios, offline data transfer can make sense, and Microsoft offers Azure Data Box Disk solutions for this purpose.

You can order the Data Box through the Azure Portal. Microsoft ships the Data Box to your site where you upload data through a local web user interface. You then ship it back to a Microsoft Azure Data Center, and the data is uploaded there into your Azure Storage account. You can track this process through the Azure Portal.

The Data Box can be used for one-time migration of very large data quantities. If data is subsequently gathered and needs to be uploaded to Azure, the incremental volumes for updates are usually a fraction of the size and can likely be handled by network connections to Azure. If remote locations periodically become disconnected and very large data volumes are again gathered, a Data Box can be requested again and used to move these larger data volumes to Azure.

The standard Azure Data Box comes in three variations: Data Box Disk (40 TB of disk, 35 TB of usable space), Data Box standard (100 TB of disk, 80 TB of usable space), and Data Box Heavy (1000 TB of disk, 800 TB of usable space). You can also send up to ten of your own disks to an Azure data center for loading. Figure 7-8 illustrates the choices of Azure Data Box selections viewed through the Azure Portal.

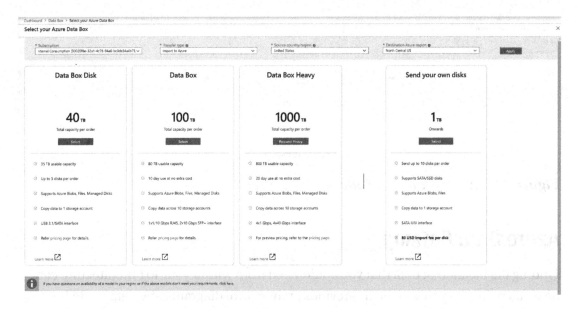

Figure 7-8. *Azure Data Box creation*

Alternatively, you might choose online data transfer. Data Box Gateway is a virtual device residing on-premises supporting NFS and SMB as protocols. The virtual device transfers data to Azure block blobs, Azure page blobs, or Azure Files.

Azure Data Box Edge is a physical device with the gateway capabilities of the Data Box Gateway. It additionally can be used to preprocess data including performing data aggregation, modifying data (e.g., removing sensitive data), selecting and sending only subsets of data, or analyzing and reacting to IoT events locally. You can apply ML models at the edge before data is transferred.

For example, the Data Box Edge can be used to capture video and, configured with the Azure IoT Edge Runtime, can push video frames through a FPGA-based AI model. The source code for such an application is posted on GitHub.

Figure 7-9 illustrates the Azure Portal interface used in creating the Data Box Gateway and Azure Data Box Edge.

Figure 7-9. *Azure Data Box Gateway/Edge creation*

Azure Data Catalog

Finding the location of data that you could need from within your IoT architecture can be challenging. As you've seen in previous architecture diagrams, the data could be stored in a variety of locations and in a variety of types of data management systems.

Azure Data Catalog is a cloud-based service where data management locations in your implementation can be registered and metadata describing the data in each location is stored. The metadata is searchable and can be enhanced by users of the Data Catalog, enabling crowdsourcing of descriptions, tags, and other descriptive metadata. The search capability makes finding where data is located much simpler.

Typical metadata includes

- Name of asset

- Type of asset

- Description of asset

- Names of attributes or columns

- Data types of attributes or columns

- Descriptions of attributes or columns

Figure 7-10 illustrates an interface in the first-generation version of Azure Data Catalog. We searched for data tagged as "sensor." On the first page of search results, we found that sensor data is stored in SQL Data Warehouses, HIVE databases and tables, and an Azure container. When we select one of the SQL Data Warehouse locations, we see information on filters available, experts assigned, glossary terms, and user tags that have been assigned to the sensor facts table on the left side of the interface. On the right side, we can provide a friendly name, better description, add experts by providing their email addresses, add tags, change connection information, and can take ownership of management.

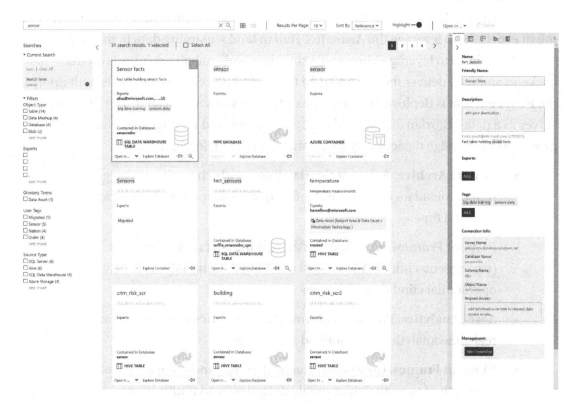

Figure 7-10. *Azure Data Catalog view of sensor data management locations*

Through the interface, we can also open the tables through a variety of interfaces. For the sensor facts table, we can browse the data contained in the table in Excel (all or just the first 1000 entries) or use SQL Server Data Tools, Power BI Desktop, or Power Query. If we choose to explore a database, we see all the tables present in the database and can explore each one.

Data Historians and Integration to Azure

Vendors of earlier-generation IoT solutions designed for entirely on-premises deployment of those solutions. They provided landing spots for device data in on-premises time series data stores called data historians. Today, many of them are moving key components to the cloud. The components they choose to move vary from vendor to vendor. Here, we'll look at how OSIsoft and PTC ThingWorx are leveraging Azure. As you might expect, both leverage the Azure IoT Hub to land streaming data in the cloud.

OSIsoft chose to leverage Microsoft Azure data and analytics solutions on the backend and fully deliver their platform at the edge. The OSIsoft solution is built upon the PI System that is deployed on-premises where devices are located. The PI Server serves as a data historian as it is used to capture, store, and manage data that is being produced by the edge devices. Key components in the PI Server include

- **PI Data Archive.** A time series database in which the data is tagged so that metadata can also be used to query other data besides the date and time.

- **PI Asset Framework.** Maps sensor readings in tags into models (e.g., parent–child relationships) enabling determination of the operational condition of devices.

- **Asset Analytics.** Enables real-time device metrics to be viewed and real-time analytics to be applied at the edge.

- **PI Event Frames.** Conditions can be defined producing events of interest tied to analytics, anomaly detection, and notifications.

- **PI Notifications.** Enables alerting of operators when defined thresholds are met or exceeded, or failures or other anomalies occur.

Integration from the PI System to Azure for streaming data is provided by the PI Integrator Advanced Edition that serves as a gateway to the Azure IoT Hub. Integration from the PI System to Azure is provided by the PI Integrator Standard Edition (for batch feeds) directly into Azure SQL Database or Azure SQL Data Warehouse. These are pictured in Figure 7-11.

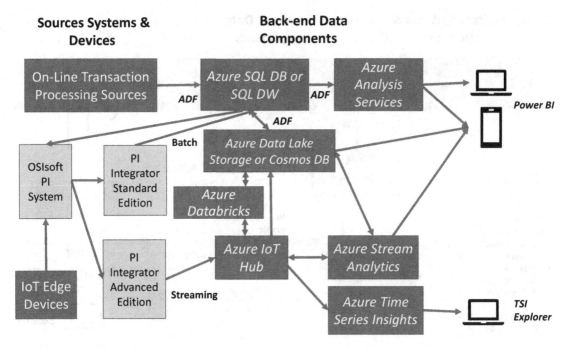

Figure 7-11. *OSIsoft PI System integration with Azure components*

Note the similarity on the backend to previously described Azure architectures. PI Vision provides an additional mobile/desktop interface in the OSIsoft solution.

In comparison, PTC chose to develop ThingWorx with components deployed across the entire ecosystem. Within the IoT Edge, PTC provides Software Content Management (ThingWorx SCM), Remote Access and Control (ThingWorx RAC), and Industrial Connectivity (Kepware). Microsoft's complementary offerings include Azure Stream Analytics, Azure ML, and Azure Functions in containers.

In the backend, the ThingWorx historian is hosted on Azure PostgreSQL. PTC also provides an Asset Advisor and machine learning (ML) solution. Microsoft provides the Azure IoT Hub, Blob Storage, Time Series Insights, Azure Active Directory, and Azure Machine Learning as important complementary components.

Figure 7-12 illustrates how all these components fit together in a reference architecture.

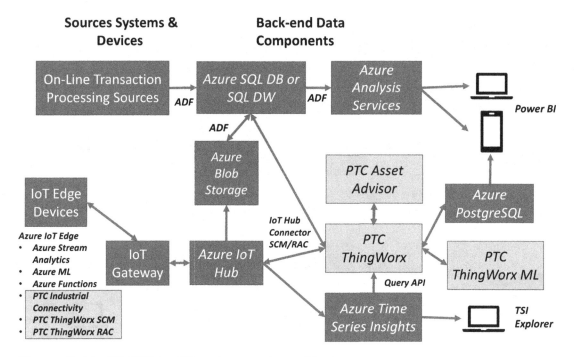

Figure 7-12. *PTC ThingWorx integration with Azure components*

Not all the components pictured in the two previous architectures are always implemented. The historians for each of these companies are widely present in their installed bases, so the components directly tied to the historians tend to have the greatest adoption.

CHAPTER 8

Developing a Plan for Success

Throughout the previous chapters in this book, we've focused primarily on the technical aspects of designing and deploying a Microsoft Azure-based IoT architecture solution. But we also touched on business aspects as we described some of the potential use cases. By now, you might still be wondering how to determine where IoT will be most beneficial in your organization and how to build support for such projects.

Defining and gaining sponsorship for these projects often incorporate "design thinking," a methodology that evolved since the early 1990s to become widely adopted in innovative technology projects like IoT initiatives. In 1992, Richard Buchanan connected design thinking to innovation in a work titled "Wicked Problems in Design Thinking." David M. Kelly founded the design consultancy named IDEO at about the same time and based its processes on design thinking concepts. Today, many major universities teach design thinking in curriculum focused on techniques used to drive innovation.

In this chapter, we'll cover identifying the right initiatives and dig deeper into techniques used to prioritize projects and maintain an agile approach to problem-solving, prototype creation, and testing. We also cover building support for the project as you move from prototype creation and testing to full-scale implementation.

The major sections of this chapter are

- Identifying the right initiatives

- Moving from prototypes to implementation

- Some final thoughts

As we cover identifying the right initiatives, we'll describe how the design thinking approach can help you during this stage in your project.

163

R. Stackowiak, *Azure Internet of Things Revealed*, https://doi.org/10.1007/978-1-4842-5470-7_8

Tip   If design thinking is a new approach to you, we hope that you'll gain a basic understanding of the approach by reading this chapter. Design thinking is widely used today in a variety of technology projects, especially in innovative software development and deployment. In addition to IoT projects, you'll likely find it practiced in the development of artificial intelligence applications and blockchain-based solutions. You might consider taking online or in-person courses to augment your own understanding given the popularity of design thinking as an approach in many technology areas.

Identifying the Right Initiatives

Innovators and lean startups commonly develop solutions in a sequence of events that consist of hypothesizing, designing, testing, and learning. These are fundamental activities present in the design thinking approach. Design thinking can be defined as a series of steps that include

- Observation and research

- Problem definition

- Ideation

- Prototype creation

- Testing

- Implementation

Identifying the right initiatives and solutions to build usually consists of the first five of these steps, beginning with observation and research. This first step seeks to gain an empathetic understanding as to how work is performed and the challenges that are present. It is a prelude to identifying problems and potential opportunities to do this work better in new and different ways. Workers are interviewed or observed, and sometimes developers who will be assigned to the project are also immersed into the experience.

Other research can be initiated that gathers information on internal corporate goals and initiatives, similar initiatives that are in progress at competing companies, and emerging trends in the industry at large. Some sources of places to gather this intelligence include financial earnings statements, presentations provided by companies to investors, trends and case studies described in industry trade journals, and presentations made by experts and insiders at industry trade conferences.

The problems that exist are then defined and framed. We consider the points of view of various potential stakeholders regarding the nature and scope of these problems. Our previous findings can help us create a list of compelling needs and problems that will fuel our brainstorming attempts to identify potentially innovative solutions.

During the ideate phase, we might use a variety of techniques to identify potential solutions and evaluate them. We seek a wide diversity of ideas in problem-solving. We also begin to prioritize which ideas are worthy of prototype development.

The creation of prototypes makes solutions tangible for stakeholders. The development of storyboards, other visuals, or physical builds using technology components are pursued. Multiple solutions to the same problem might be tested, explored, compared, and refined. An important goal is to succeed or fail inexpensively and quickly during the prototype and testing phases.

When testing occurs, stakeholders (including users of the solution) evaluate prototypes and provide their feedback. Problems become better framed, and the most likely viable solutions become better understood. If testing proves successful, we might then move forward with a full-scale implementation.

Figure 8-1 illustrates this cycle following the outer ring of arrows pictured. The arrows in the center remind us that we might need to return to previous steps. For example, when we observe how our prototype functions in testing, we might decide that doing further research and redefining the problem might be necessary.

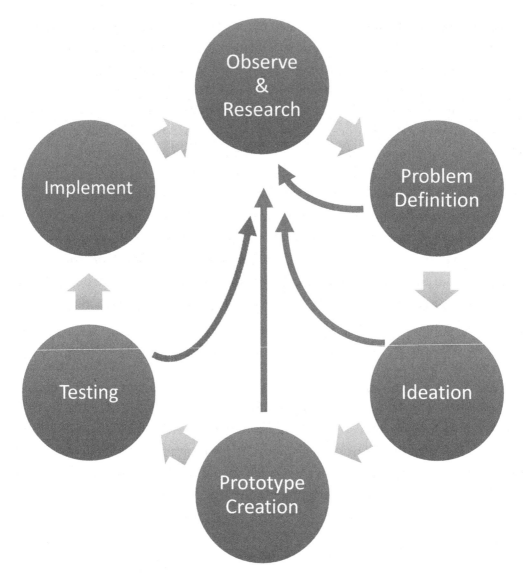

Figure 8-1. *Phases in design thinking*

Implementation of a production-ready solution occurs once we've determined that we are ready to put the proposed solution into full-scale operation. The implementation phase requires additional planning and designs for reliability, availability, serviceability, and security. Often, a pilot or operational prototype is developed at smaller scale with these requirements in mind to get a better picture as to the true cost of the final solution. A pilot can also help validate technical feasibility. A roadmap to full implementation might be required with identification of proposed costs and estimated value of the solution at various steps along the way in order to gain budget approval.

Next, we'll take a deeper dive into each of these phases with practical guidance on how to execute each phase.

Observe and Research

We begin to determine the problems that need to be solved by observing top-of-mind challenges within the lines of business and opportunities for success. We also explore emerging threats external to our company or organization. At this first stage, we are looking for potential stakeholders for projects who possess visions of what the future might look like.

To get our arms around the state of business processes, we document the current environment and interactions that take place. As we do this, we seek to identify opportunities to improve efficiencies, quality of goods, quality of services, quality of production, and/or safety. We also analyze any tools and technologies used in these processes and document how workers use information and respond. We begin to understand the importance of such tools and technologies in successfully executing necessary business processes and uncover any vested interest or reluctance to change that the users might have.

Some of the typical information we might gather and document through interviews with users where we suspect an IoT initiative could emerge includes

- A description of normal activities and situations that users encounter

- A description of abnormal activities and situations that the users encounter including the frequency of abnormalities and the impact on the business

- A description of feedback from systems in response to normal situations

- A description of feedback from systems in response to abnormal situations

- A description of how users know what appropriate actions to take during abnormal situations

- Critiques about the amount and level of detail in information that is provided

- Critiques about the complexity of response required in abnormal situations

- Suggestions for process and system improvement

In addition to interviewing individuals, we might simply watch their actions in normal and abnormal situations, including their interactions with co-workers and systems. We might also capture these interactions through video recordings that then can be used for further study and to augment the written record of our observations.

When we document the problem solution process, including current activities and outcomes, we can present the process in the form of a journey model. A typical model that might lead to an IoT project begins with a description of how workers are informed of the status of a specific process and the problems and abnormal situations that might impact that process. Workers then decide what appropriate action is necessary if a problem is confirmed, perform some sort of action to fix the problem, and see a response confirming that the remediation action has occurred.

As an example, let's look at a worker monitoring the soldering of components to a circuit board on a production line. They visually determine if the components are in or out of alignment. Their action could include stopping the production line, determining where the misalignment is occurring, correcting the cause of misalignment, and restarting the production line. The response would then be a validation of the return to normal production. Figure 8-2 shows how we might illustrate this model.

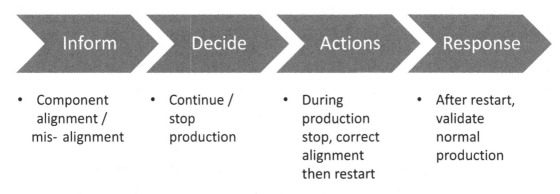

Figure 8-2. *Example journey model*

If the focus is only on improving existing internal processes and/or meeting current needs, we might lose sight of innovations that are happening elsewhere. Interviews with line of business leadership are often a great place to gain a better understanding as to what competitors are up to since having a current understanding of innovative initiatives in their industry is usually part of their day job.

To gain a more complete understanding of external influencing factors, a PESTEL analysis might be performed. Such an analysis weighs political, economic, social, technological, environmental, and legal factors. These factors can drive increased momentum toward more automated IoT solutions.

Relevant political factors helping to provide momentum might include tax incentives for modernization. Economic factors in play could include consideration of increased labor costs and the need to maintain or grow margins. Social factors might be driving the workforce to gain technical skills that rely on IoT-based solutions. Technological advancements could be driving a need for faster innovation to stay competitive. Environmental factors could include demands for sustainability and the need to reduce waste in processes. Legal factors might include the introduction of new regulations and laws related to safety and products.

You might determine that some independent research is in order, especially as competitors begin to enter new business areas and PESTEL-related factors influence the need for IoT solutions. Where can you find such information?

Start with quarterly financial statements and presentations to investors. Pay attention to IoT-related vision statements by leading executives and how they respond to business analysts' questions on earnings calls that could lead to such projects. Be on the lookout for statements regarding the impact of competition and PESTEL-related drivers that are discussed during these calls.

You should also research industry and government sources of information that can provide insight into additional drivers. Pay special attention to conferences where other similar companies and organizations are speaking about their IoT initiatives and attend those conferences. Though technology vendors sometimes present compelling use cases, presentations of initiatives by business stakeholders within similar companies and organizations to your own can provide you with a great deal more insight into challenges in implementation and the true nature of the business drivers and benefits obtained. Attending such industry and technical conferences can help you determine where you should focus your own initiatives.

Problem Definition

In the previous phase, we began to uncover problems that could be worth solving and that could drive IoT initiatives. In this phase, we better define these problems and look at them through various points of view. We'll also begin to understand what each group sees as benefits in solving these problems. And we begin to prioritize the importance of solving the defined problems.

Problems solved by IoT initiatives can impact many different stakeholders beyond the frontline workers. Business sponsors likely have their own unique set of challenges and objectives. Senior leadership in the company or organization and their partners might each have unique views regarding challenges and the problems that need to be solved. Figure 8-3 represents various points of view that could be present.

Figure 8-3. *Points of view impacting problem definition*

For each targeted group, we should fully understand their business goals and any tasks and solutions in place to achieve those goals today. As we gain that understanding, we should document limitations and inefficiencies that are present. We will also want to gain an understanding of each group's desire for change and any alternative solutions currently under consideration.

In some situations, we might uncover problems that are exacerbated by inadequate skills or extraordinary physical effort required to overcome current solution shortcomings. Though training might overcome some of these problems, we should also be on the lookout for unusual worker turnover where additional training has not solved problems.

For each group, the perceived value and range of benefits in solving a problem could differ. For example, some might see a problem as one of efficiency, while others might see the same problem as a quality improvement problem. This could lead to a divergence on the vision of what ideal solutions might look like. So, the input from each targeted group might include not only broad benefits from improvement in business processes that solve a specific problem but also benefits that will be personal to each target.

As we gather a list of problems, it is unlikely that we will have unlimited resources to solve all of them. So, we should keep in mind that some prioritization of the order in which we solve them will be needed. We should start to gather notes that will help us understand:

- Potential return on investment (ROI)

- Time to return on investment/solution

- Importance to C-level business leaders

- Regulatory requirements

- Cost

- Risk of deployment in deploying a new solution

- Risk of not deploying a new solution

- Skills needed/lacking

- Culture alignment of potential solution

- Device sophistication and availability of needed quality data

You might think that in most organizations only consideration of potential return on investment drives the selection of the most important problems to focus on. However, this is frequently not the case. The time that it takes to get to a viable solution and a positive return on investment can be the determining factor when the window of opportunity for solving the business problem is short, there is a very limited budget, management is exerting pressure to solve the problem, and/or the problem is seen as an extremely dangerous competitive threat.

Sometimes, problems must be solved regardless of the ROI. For example, C-level executives might have issued forward-looking statements to investors that promise delivery of a business solution built upon IoT. Regulatory requirements might also drive the need for creation of an IoT solution to address mandates present in some industries.

Several of these factors might cause us to deprioritize certain initiatives. A project might be viewed as too costly or risky, regardless of the ROI that might be obtained. Skills could be lacking to implement or use the proposed solution, and the culture in the organization might not be ready for adoption of the technology or the solution. There could be issues regarding availability of quality data due to a lack of devices, sensors, or other infrastructure.

In many companies and organizations, multiple considerations in combination impact the determination of priorities in project funding decisions. Identifying that mix of prioritization considerations during this phase will be critical in helping you to determine which problems to focus on solving.

Table 8-1 illustrates the capturing and prioritization of three potential IoT initiatives in a manufacturer that we will use as an example. We've identified key stakeholders for each initiative, the important metrics required in each proposed effort, noteworthy data considerations, and the potential business impact. We have also assigned a priority to each initiative.

Table 8-1. *Capturing and prioritization of initiatives*

Initiative	Stakeholder(s)	Metrics Required	Data Status	Business Impact	Priority
Minimize downtime	VP Manufacturing	Uptime, line rate, operating conditions	Need additional sensors on lines	On-time delivery, increased revenue	1
Minimize rework	VP manufacturing, VP quality	Accepted/rejected products	Need additional sensors, data quality issues	Decrease cost of goods sold by optimizing manufacturing process	2
Minimize warranty claims	VP quality	Production line, rate of return	Gather production lines and worker data	Decrease set-asides covering warranty expense	3

In this example, the company is prioritizing efforts that will increase revenue. Cost containment might also be important but is of secondary concern. So, minimizing downtime on the production lines is listed as the top priority, though we'll likely need to add sensors to the lines or purchase new equipment to gather the metrics that we'll require.

Ideation

Ideation is the start of solving the identified problem or problems in the initiatives. In IoT initiatives, a goal of this phase is to gather many solution ideas and determine a solution worthy of developing a prototype that will then be tested for validity in front of stakeholders and interested parties.

A variety of techniques are commonly used during the ideation phase. Often, the ideas are generated through facilitated discussions. Brainstorming techniques can be used to bring about a free expression of ideas. The focus is on gathering a large quantity of diverse ideas. No criticism of an idea or idea ownership is allowed. Group members in the exercise are ideally a heterogeneous mix of individuals (not just experts). Everything is written down and captured.

The facilitator has the important role of guiding the discussion. They might occasionally solicit different points of view intended to drive consideration of new and diverse ideas. A best facilitation practice is to ask open-ended questions. The session might begin with the question, "How might we solve the defined problem?" A question sometimes asked to spur insightful discussions is, "How might our toughest competitor solve the same problem?"

Using our earlier prioritized initiatives from our manufacturer example, we would begin by soliciting input on how we might minimize downtime on our production line. During brainstorming, ideas on how to solve this problem can be written by team members on Post-it Notes of different colors, or they might use software-based applications featuring the equivalent of Post-it Notes to share their ideas. Each team member has a uniquely colored notepad so that we can determine where the ideas are coming from (to encourage broad participation) and so that team members can track their ideas as solutions areas are defined.

A sample of some of the ideas that might be gathered in our example scenario includes video training of workers, rotation to different lines during shifts to prevent boredom, automated gathering of assembly line speed statistics, measurement of increased vibrations or abnormal variations in speed caused by equipment problems, and better understanding of bill of materials including quantities on-hand for goods production.

The facilitator will see that these and other ideas gathered fit into broad solution themes. Solution themes in our example include training of personnel, staffing model changes, better capture of metrics measuring output, earlier indication of potential production line problems, and earlier indication of supply chain shortages. Figure 8-4 illustrates how the Post-it Notes are aligned into these solution themes.

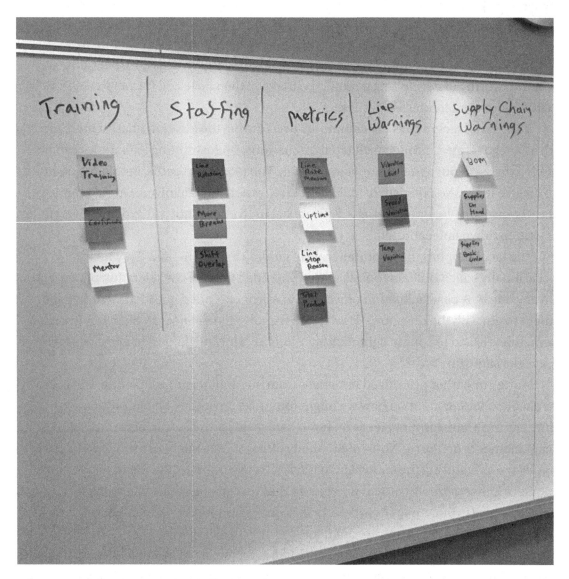

Figure 8-4. *Brainstorming solutions using Post-it Notes*

The content on the boards containing the sticky Post-it Notes is usually captured by taking pictures of the results with a mobile phone camera (if an electronic whiteboard application that can capture these results wasn't used instead).

Note   The application software used for ideation that features the equivalent of Post-it Notes has several advantages. Participants can be in remote locations. Visibility into the notes being written occurs much faster, and deduplication of identical ideas is easier. Such applications can also enable faster classification of ideas and voting among participants.

The output of this exercise is later documented in a tabular format such as that illustrated in Table 8-2.

Table 8-2. *Brainstorming solutions to minimizing production downtime*

Solution Themes	Post-it Note Ideas	Solution Votes
Personnel training	Video training, certification, mentoring by supervisor sessions	
Staffing model changes	Line rotation during shifts, more breaks, more overlap during shift change	
Automated output metrics measurement	Line rate, uptime, line stoppage reason, total products produced	
Production line warnings	Vibration levels, speed variation, temperature variation	
Supply chain shortage warnings	Bill of materials/production planning, supplies on-hand, supplies backordered	

Some of these ideas and solution themes shown here will not lead us toward pursuing an IoT project. That's ok as the primary goal is to solve the problem at hand, not force fit a technology solution. A vote is taken among the participants regarding the most important solution themes, and that will guide us regarding the next steps to pursue. If IoT projects don't make the cut, we've just determined that funding for developing and putting an IoT solution into production later could be unlikely to occur anyways.

This brainstorming technique used during ideation is also used during other phases when design thinking techniques are used in an agile sprint. We discuss the agile sprint later in the chapter.

Tip   Should all votes count the same? Should everyone vote? At this stage in ideation, usually every vote does count the same and everyone votes. If these alternatives are reconsidered later, you might then give more weight to key stakeholders who can fund the project and the potential users. Prior to this later vote or discussion, you might decide it beneficial to create an influence map that shows which individuals are supporters, neutral parties, and against solving the problem. You'll also want to note whether they are responsible, accountable, consulted, or simply informed when it comes to solving the problem. Those who are responsible or accountable have the most skin in the game, and their views are particularly important when you determine where to narrow the focus.

Further evaluation of these solution themes generally occurs before the type and scope of a prototype are determined. Proposed solutions are typically broken down into their process steps. Variants in how those processes will be executed are explored. We also document the necessary resources that are available and the resources that we must add for the project to be viable (or indicate work-arounds that might also prove to be adequate).

Let's assume that "Automated Output Metrics Measurement" received the most votes among the team. Upon further exploration, we see that the way in which needed information is gathered today is through manual input of data into spreadsheets by production line managers. The total products produced is gathered from a counter at the end of each shift. Data regarding uptime and line stoppage reasons is manually input and subject to error based on the skills and attentiveness of the production line manager.

We've theorized that a more automated approach to gathering this data will help us increase production since we will be able to fix problems that are occurring much faster. We'll also eliminate some of the waste that currently occurs because we throw away many of the products produced during production line problems.

As we evaluate the technical capabilities needed, we might decide that additional sensors can help us better measure what is really happening. Alternatively, or additionally, we might also determine that there is an opportunity to introduce cameras and use image recognition on the production line to monitor the line and trigger more immediate actions when needed.

At this point, we probably would revisit this solution idea with the team. We might present pros and cons of deploying and utilizing the various solution resources being considered as shown in Table 8-3.

Table 8-3. *Automated Output Metrics Measurement alternatives*

Solution Resources	Pros	Cons
Automatic counter, manual entry (current)	No additional infrastructure, training	Inaccurate stoppage times and reasons
Add only sensors to production line	Gathers actual stoppage time and reasons more accurately	Cost of equipment retrofit, software development
Add only image recognition to production line	More accurate than manual observations over time, potentially more immediate reaction to production problems	Cost of cameras and software. Need negotiation with union?
Add combination of sensors and image recognition	Potentially the most accurate, also key to enabling production line warnings	Cost of equipment retrofit, cameras, and software. Need negotiation with union?

As documented here, the pros and cons of a solution idea for solving a problem is being evaluated for technical feasibility, user desirability, and business viability. Other factors such as adaptability, sustainability, and scalability might also be evaluated.

We might also decide to evaluate each alternative by using more perspectives than simply pros and cons. One way to do this is by evaluating the strengths, weaknesses, opportunities, and threats associated with each alternative in what is often referred to as a SWOT analysis.

We'll now compare a SWOT for simply adding sensors providing Automated Output Metrics Measurement to a SWOT for adding both sensors and image recognition as means to solve our problem. We'll begin with the SWOT for the adding sensors alone alternative in Table 8-4.

Table 8-4. *SWOT for adding sensors alone in Automated Output Metrics*
Measurement alternative

Strengths	Weaknesses
• Gather actual stoppage time	• Cost of equipment retrofit over status quo
• Automated gathering of reasons for stoppage through sensors (more accurate)	• Cost of software development over status quo
• A good first step toward improving production	• Still heavily dependent on manual inspection
Threats	**Opportunities**
• Skills needed to build and maintain	• Develop new skills
• Lots of old equipment present	• Replace old equipment with modern equipment
• Competitors are updating their plants with modern equipment	• Begin to set the stage of proactive management of line

You can see that we've called out reasons why we might want to go forward with this alternative in the "Strengths" quadrant describing positive immediate outcomes from successful deployment and the "Opportunities" quadrant describing positive longer-term impacts to the company. But we've also called out reasons we might not want to go forward with this alternative in the "Weaknesses" quadrant by documenting the perceived shortcomings of the approach and in the "Threats" quadrant documenting the challenges that could impede the project and limit its success.

The alternative that includes both sensors and image recognition in the solution introduces some different strengths, weaknesses, opportunities, and threats. Table 8-5 illustrates a SWOT for this alternative.

Table 8-5. *SWOT for adding sensors and image recognition in Automated Output Metrics Measurement alternative*

Strengths	Weaknesses
• Gather actual stoppage time	• Cost of equipment retrofit over status quo
• Automated gathering of reasons through combination of sensors and images (most accurate alternative)	• Cost of hardened cameras over status quo
	• Cost and complexity of software development over status quo
Threats	**Opportunities**
• Skills needed to build and maintain	• Develop new skills
• Lots of old equipment	• Replace old equipment with modern
• Possible union challenge regarding cameras on line	• Strongest alternative that sets the stage of proactive management of production line

We now can compare the relative strengths, weaknesses, opportunities, and threats in the two alternatives. But how do we determine which alternative is the best one for our situation?

Alternative approaches to solving the same problem are often evaluated using agreed upon criteria for comparative scoring. Examples of such criteria can include

- Strategic importance

- Competitive importance

- Feasibility

- Return on investment

- Time to working prototype and testing

- Time to production and return on investment

As an example, we decided to score the two Automated Output Metrics Measurement alternatives compared in the previous two SWOT tables. Table 8-6 illustrates these alternatives scored using these criteria on a scale of least value or likelihood of optimally occurring (1) to most value or likelihood of optimally occurring (5).

Table 8-6. *Scoring of Automated Output Metrics Measurement alternatives*

Alternative	Strategic Value	Compete Value	Feasible	ROI	Time to Test	Time to ROI	Total
Adding sensors alone	3	4	4	4	4	4	23
Adding sensors and image recognition	5	5	4	4	3	4	25

The better understanding that one has of the trade-offs, the more likely one is to pick the right prototype creation strategy. Based on the scoring recorded in this table in which we see that one of the alternatives received a higher score, we would likely choose to proceed with the option that both adds sensors and image recognition to the production line.

Prototype Creation

The purpose of prototype creation is to enable testing of the validity of ideas gathered in the previous phase. At this point, we have hypotheses about our potential solution(s) to the problem that we identified. Prototypes can come in a variety of types and sophistication, with the cost and sophistication of prototype creation generally aligned to the degree of commitment to solving the identified problem.

Innovators use prototypes to experiment as they formulate solutions. They expect failures but continually learn through prototyping in the most cost-effective manner possible. They see solution development as an evolutionary process but are not afraid to throw away efforts that early-on prove to be extremely difficult to implement or are impractical in other ways.

Prototype creation sometimes goes through a series of stages. An initial stage might be the creation of a storyboard describing how and what the solution to the identified problem will deliver. How the solution will change business processes might be defined. Needed functional components in the solution are identified.

Mock-ups of the functional components can be created once there is agreement regarding the definitions in the storyboards. These might include versions of the visual interfaces that will be provided. Some of the available technical components might also be used for functional illustration.

A next step could be a technical proof of concept. In this step, we further identify data and integration challenges as well as skills gaps. Up to this prototype development phase, throwing away portions of previous efforts might have had little consequence in the overall cost of creating the solution. At this point, we begin to make a more significant technical investment with the notion that we'll continue to evolve this prototype over time.

Our technical prototype can be used to demonstrate what the final solution could look like from a functional standpoint. Some also use this effort to determine the operational impact and other potential gaps in the proposed solution.

Figure 8-5 illustrates a typical series of steps in prototype creation.

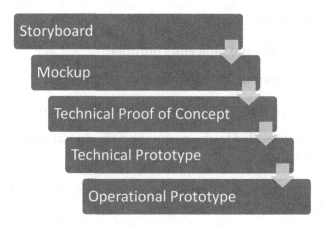

Figure 8-5. *Typical prototype creation steps*

Testing

During each stage in prototype development, testing is used to solicit feedback from stakeholders and other key parties, especially the frontline users of the proposed solution. We use testing to validate our hypotheses about the value that our solution will deliver. Going into testing, we should have criteria established that define the outcomes that we are expecting to see if our hypotheses are correct.

In early stages of prototype development, stakeholders and key parties view storyboards and mock-ups. They are then interviewed and/or share their observations in discussions or surveys. To get broader feedback, focus groups are sometimes created. When there is a lack of consensus, votes might be taken on components and the overall solutions to assure that choices are made that will have the broadest support.

Testing can help identify the must-have components of the solution since the lack of any such components would be pointed out in the feedback received. Satisfaction that the solution requirements are being met will influence the participants' views on quality of the effort made by the team and of the solution itself. Features added that go beyond the basic requirements might improve perception of the solution or might have little impact (other than adding cost). It is important to document all of this in the feedback and capture suggestions for improvements.

Questions are often formulated such that answers can be provided on a sliding scale. For example, when testing the Automated Output Metrics Measurement solution prototype for monitoring the production line, participants might be asked to rate the following on a scale of 1 to 10, where 1 is poor and 10 is excellent:

- Ease of use of the prototype

- Clarity of messages and indicators provided

- Clarity of directions on how operators should respond

- Timeliness of messages that appear and quality of automated actions

- The response provided by the prototype if the operator makes errors in fixing problems

- The prototype's potential to teach new operators how to more effectively do their jobs

- Technical feasibility of the prototype including apparent reliability, availability, and serviceability

- Overall satisfaction of operators/workers with the prototype

During the prototype development phase, organizations will sometimes have multiple teams creating prototypes in competition with each other. This can provide a means to get to a more comprehensive solution faster. In our earlier example, one team might be adding sensors alone to a production line and building out that prototype. A second team could be adding sensors and image recognition to a second production line.

Testing and evaluation of competing prototypes provide us with comparisons of the solutions but also enable us to judge the quality of work by each team. One prototype might be chosen over another if it meets all key requirements and is perceived to be the best solution or is more cost-effective. Sometimes, the best features or components from each prototype are determined and then merged into a single solution.

The Agile Sprint Approach

Agile sprints have become a popular technique used in early prototyping and testing of potential software and applications improvements. The sprints generally take place over short periods of time (2 to 4 weeks) with requirements driven and prototypes reviewed by a small group of stakeholders and interested parties. Each iteration of the entire process previously described in this chapter is compressed into this short time frame.

In preparation for the sprint, the facilitator does research into potential problems that might be addressed, observes current solution practices, and determines who the likely stakeholders and interested parties will be. The sprint begins with the facilitator leading this group of participants through the initial problem definition and ideation phases. A diverse group of six to ten people takes part.

Just as in longer engagements, the facilitator briefs the group on the openness of the process and reminds them not to feel limited by constraints and to set aside any critiques. Both the problem definition and ideation phases often utilize the same brainstorming techniques. These phases commonly take place on back-to-back days. The same group of participants is usually present for both.

At this point, a prototype is started and developed, usually over no more than a 2-week or 3-week period. The prototype is then "tested" in front of the participants to validate whether it might provide the solution that the participants were looking for. This discussion can drive further iterations of this process or lead to a decision to move forward with an implementation.

In 2005, this process that we describe here was summarized in a "Double-Diamond" figure by the Design Council UK. The Double-Diamond visually represents where the scope of what is being considered widens and where it narrows. Figure 8-6 is our representation of the Double-Diamond using the nomenclature we've used in this book.

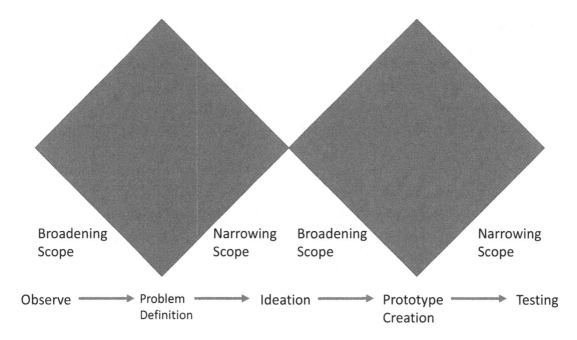

| Broadening Scope | | Narrowing Scope | Broadening Scope | | Narrowing Scope |

Observe ⟶ Problem Definition ⟶ Ideation ⟶ Prototype Creation ⟶ Testing

Figure 8-6. *The Double-Diamond design process in an agile sprint*

During observation and research, the scope of our effort widens. As we define potential problems to solve, we eventually narrow the scope, often to a single problem. As we gather ideas about solving that problem, our scope widens again. In the prototype creation and then testing, our scope is narrowed, eventually to a single solution as we move toward a production implementation.

Regardless of whether the agile sprint drives prototype creation and testing, or design thinking drives it at a slower rate, funding for a full implementation and operationalizing the solution will require additional considerations. We discuss these in the next section of this chapter.

Moving from Prototypes to Implementation

At this point, our prototypes have been tested, and we have been listening to our constituents. We have aligned our prototype functionality with desired business goals and made modifications where needed.

We now we likely have more questions that must be answered before we can move toward deployment of a production environment. These questions can include

- Is there measurable return on investment or significant business value from the solution when we operationalize it, and when will we see it?

- Can we operationalize the solution in a reliable, manageable, serviceable, and secure manner?

- Who should implement the production version, and how should it be implemented?

- What sort of roadmap will assure sponsors and stakeholders such that the project receives adequate funding needed to deliver the production version?

Let's explore how we might answer these questions.

Measurable Return on Investment

We noted earlier in this chapter that return on investment is just one of the considerations used in determining which projects move forward. That said, once projects move beyond the testing and prototype phase, ROI analysis frequently becomes necessary to justify making the investment in an operational version of the solution. This is particularly true where the CFO has a role in approving these projects.

ROI is computed over a time period that includes developing the operational solution and then the subsequent period when the solution is in operation. ROI is positive when the business value provided exceeds the total cost of ownership (TCO) over a given time period. A goal in these projects is usually to reach positive ROI as soon as possible. The formula for ROI can be expressed as

ROI = (Business Value – TCO) / TCO

The TCO components in a typical IoT project can include

- Azure subscription costs of relevant backend components

- Development and deployment of IoT Edge, IoT Hub, data management, and analytics/machine learning software solutions

- Integration of IoT components

- Integration to legacy components (where required)

- Internal staffing supporting the Azure deployment

- IoT device and networking purchase costs (including upgrading of legacy equipment where required) and installation costs

- Ongoing IoT device and networking support and maintenance costs

- Training of internal technical staff

- Training of IoT solution operators of equipment

Typical measured business value comes from

- Increased revenue from existing and new products and business services

- Optimization of limited resources, facilities, and/or supply chain

- Improved quality of products and services

- Savings from reduced unwanted and unplanned equipment downtime

- Elimination of risk, regulatory penalties, and need to set aside monies for other related expenses

- Improved safety resulting in savings from minimized lost worker time, reduced workman's compensation, and reduced healthcare expenses

The potential business value from an IoT solution might initially be viewed by some in an organization as speculative prior to the deployment of the operational system. Such estimates of business value are most believable when coming from responsible business leadership. Often, a range of estimates is provided that can be described as conservative, pragmatic, and aggressive with pragmatic being considered the most likely scenario based on business judgement.

A typical illustration of when ROI occurs over time in an IoT project is shown in Figure 8-7. Initially, the cost of development and TCO far exceeds business value. Later, TCO becomes primarily support-related and bringing additional IoT devices online, while business value continues to grow. The crossover point where ROI is positive (business value exceeds TCO) is sooner when using aggressive business value estimates and later when conservative estimates are used.

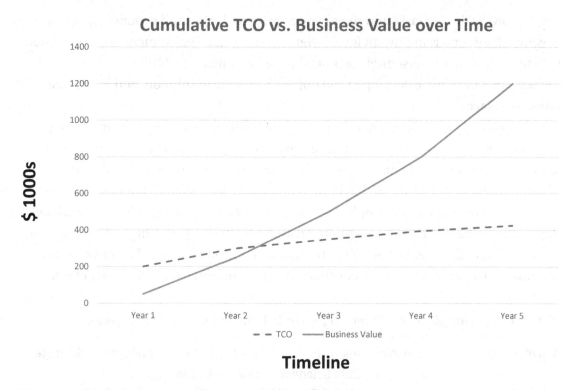

Figure 8-7. *ROI crossover illustrated by business value and TCO over time*

In Figure 8-7, positive ROI occurs just after the start of Year 2 in the project. TCO is growing at a diminished rate, while business value growth continues to increase.

Note   The value of money changes over time. Hence, costs and business value are sometimes computed using "net present value" (NPV) formulae to provide more realistic views of ROI during a project lifetime.

Operational Considerations

In Azure-based backend cloud deployment servicing IoT solutions, operational aspects are greatly simplified compared to earlier on-premises deployment of these resources. Given that most companies were adopting a Platform as a Solutions (PaaS) strategy for IoT solutions when this book was written, let's look further at key tasks and roles of key players aligned to such a strategy.

Identifying the key tasks and roles needed prior to deploying a production version of your IoT solution is important for a variety of reasons. As we estimate costs associated with full deployment, personnel costs need to be identified. In addition, understanding the skills that need to be developed will impact cost, time to solution, and hiring that must take place.

Key tasks that must be executed when full deployment occurs include managing day-to-day operations – monitoring the infrastructure, performing change management, application release management, and performance tuning – and assuring the protection of data. Operationalizing the architecture is frequently represented by a RACI diagram.

Table 8-7 illustrates an example RACI diagram for an IoT backend in an Azure PaaS deployment. The RACI diagram is prepared and validated by gathering input from each of the key individuals as to their roles and responsibilities. The table illustrates who is responsible (R), accountable (A), consulted (C), and informed (I) for a variety of roles.

Table 8-7. *Example RACI diagram for an IoT cloud backend deployment*

Activity/Task	Stakeholder/ LOB	Analyst/ Data Scientist	Azure Admin.	Data Admin.	Developer	IT Manager
Day to day Operations	I		R			A
Monitoring	I		R			A
Change management	I	I	R	R	R	A
Application release management	C	R	I	I	R	I
Performance tuning	C	C	R	R	R/I	I
Data protection	I	R/I		R		A

Similar activities and tasks are assigned to those responsible for deployment and monitoring of IoT edge resources. Table 8-8 illustrates some of the roles that might appear in an IoT edge RACI diagram where IoT devices are deployed in facilities such as manufacturing plants, healthcare facilities, campus facilities, utility plants, or others.

Table 8-8. *Example RACI diagram for an IoT edge deployment*

Activity/Task	Stakeholder/ LOB	Analyst/ Data Scientist	Developer	Device Admin.	Network Admin.	Facility Manager
Day to day Operations	I			R	R	A
Monitoring	I			R	R	A
Change management	I	I	R	R	R	A
Application release management	C	R	R	I		I
Performance tuning	C	C	R/I	R	R	I
Data protection	I	R/I		R/I	R/I	A

Implementation Strategy

The earlier stages of the design thinking approach we described previously in this chapter should have convinced key stakeholders that there is business value in the proposed IoT solution. We should now have a good idea as to what our IoT solution will deliver. The financial and operational considerations we described in this section of the chapter will also help us in developing our implementation strategy. Even so, many risks could remain when we approach operationalizing our project, especially if this is our first IoT solution. Our implementation strategy will help us mitigate those risks.

For example, the design complexity and cost of our IoT solution could raise concerns about the potential outcome. Developing a phased approach that clearly lays out the scope of deployment within the project phases, deliverables at the end of each phase, and likely cost and business value of each phase can help to alleviate those concerns.

Within the project phases, developer and deployment skills might be required that are not widely present within the organization. Such concerns can be addressed by providing an education and hiring plan or identifying technology partners that will assist in the development and deployment during these phases.

In a highly competitive environment, competitors could be developing similar IoT solutions that match or exceed the capabilities that you have planned. Monitoring these developments and having flexibility in making some adjustments in response within the project phases could be well received by key stakeholders of the project.

Deployment of our proposed solution could also have significant impact on existing business processes. Having well-thought-out change management plans is critical to mitigating concerns about this impact. Such plans often include activities that are designed to gain support for the new processes among workers and alleviate the concern of sponsors. Training can provide education on how day jobs will be impacted and generate enthusiasm for the changes.

Preparing an Implementation Roadmap

To gain needed funding for the IoT project, you might need to sell the value and viability of the project to senior management by providing an implementation roadmap. The roadmap should contain an easily understandable message about the problem to be solved, its potential business value, and the expected time to solution that will demonstrate its value.

Within a roadmap to implementation, you will likely need to also include the following:

- An explanation as to the process used in determining which business problem(s) merited solution consideration and why this problem was given higher priority and selected for an IoT implementation

- An overview of how this IoT solution will solve the identified business problem(s)

- A timeline showing project phases, costs, and business benefits

- An overview of the current state technical architecture and how the architecture will change in its future state

- An overview of project risks and risk mitigation steps

- A description of immediate next steps upon project approval including funding needed, staffing required, planned acquisition of IoT devices and their installation, planned acquisition of additional cloud resources, and immediate training and change management activities

Multiple roadmaps are often developed to address the concerns of different audiences in different levels of detail. Business and technical roadmaps each focus on answering the questions relevant to those audiences. An executive roadmap presentation presents the information we just noted at a very high level, conveying just enough information such that the executive(s) can make an informed decision.

Some Final Thoughts

We hope that you now have gained enough knowledge to define IoT solution architectures that rely upon Microsoft Azure for providing key components. You probably realized that there is a lot to consider even before you read this book. You have many options in how you might justify such projects and in the details of the architecture that you define.

Our intention was to lay out this book in a fashion such that you could build upon the knowledge that you gained in each of the chapters. You explored

- Modern IoT architecture patterns

- Azure IoT solutions overview

- IoT devices and Azure

- Landing data in Azure

- Applying analytics, machine learning, and cognitive services in Azure

- Deploying solution accelerators and managed solutions

- Integration with legacy infrastructure

- Developing a plan for success

IoT continues to evolve. Microsoft and its partners are at the forefront in driving this evolution and are enabling new and innovative business solutions. New IoT-related standards also continue to appear while previous standards evolve, addressing areas that formerly were less well-defined or understood. That said, IoT has matured a great deal in the past few years. And waiting for the next generation solutions and standards to emerge is not an option for most organizations.

You likely read this book because you have heard so much about IoT and wanted to learn more. But you might have also read the book because you are feeling pressure from your business leadership to solve problems that would benefit by deployment of an IoT-based solution. Getting started today will help you build and develop the skills you need and start you down the road of designing and deploying solutions that can make immediate impact on the business. This book is just the beginning of gaining an understanding on how to do that.

We wish you success regardless of where you are on this journey. Microsoft IoT and the Azure platform enable the intelligent edge and the intelligent cloud required in the delivery of these valued business solutions. Successful deployment of these solutions is assuring that talented individuals skilled in designing and deploying the architecture covered in this book will be in demand for years to come.

APPENDIX

Published Sources

Industrial Internet of Things Volume G1: Reference Architecture. Industrial Internet
Consortium, IIC:PUB:G1:V1.80:20170131, January 2017.

Industrial Internet of Things Volume G4: Security Framework. Industrial Internet
Consortium, IIC:PUB:G4:V1.0:PB:20160926, September 2016.

Knapp, Jake, J. Zeratsky, B. Kowitz. *Sprint, How to Solve Big Problems and Test New
Ideas in Just Five Days.* New York, NY: Simon & Schuster, 2016.

Laney, Douglas (Gartner, Inc.). *Infonomics.* New York, NY: Bibliomotion, Inc., 2018.

Lewrick, Michael, P. Link, L. Leifer. *The Design Thinking Playbook.* Hoboken, NJ:
John Wiley & Sons, 2018.

Mueller-Roterberg, Christian. *Handbook of Design Thinking.* Independently
published, 2018.

Nath, Shyam, R. Stackowiak, C. Romano. *Architecting the Industrial Internet.*
Birmingham, UK, Packt Publishing Ltd., 2017.

NIST Special Publication 800-82. *Guide to Industrial Control Systems (ICS) Security*,
May 2015.

Schenker, Jason (The Futurist Institute). *The Robot and Automation Almanac 2019.*
Prestige Professional Publishing LLC, 2019.

Schwab, Klaus. *The Fourth Industrial Revolution.* Geneva Switzerland: World
Economic Forum, 2016.

Stackowiak, Robert, A Licht, V Mantha, and L Nagode. *Big Data and The Internet of
Things: Enterprise Architecture for a New Age.* New York, NY: Apress (Springer Media),
2015.

World Economic Forum. *Industrial Internet of Things: Unleashing the Potential of
Connected Products and Services*, January 2015.

© Robert Stackowiak 2019
R. Stackowiak, *Azure Internet of Things Revealed*, https://doi.org/10.1007/978-1-4842-5470-7_9

Microsoft Online Documentation Sources

Microsoft documentation provided important source material for many of the backend and IoT components described in this book. The documentation can be found at `https://docs.microsoft.com`.

Much of the source material listed here leads to more detailed documents. Where should your start? The "What is/What are" documents in the list point to further detail for many of the components that we described. Some of the key documents we accessed included the following (listed alphabetically by title along with most recent update date at the time we wrote the book):

About Azure Bot Service, 05/04/2019.

Azure enterprise scaffold: Prescriptive subscription governance, 09/21/2018.

Azure Event Hubs – A big data streaming platform and event ingestion service, 12/05/2018.

Azure IoT Central Architecture, 05/30/2019.

Azure IoT Edge security manager, 07/29/2018.

Azure IoT reference architecture, 01/08/2019.

Azure Machine Learning integration in Power BI, 05/30/2019.

Azure Time Series Insights explorer, 05/06/2019.

Choose a real-time analytics and streaming processing technology on Azure, 05/14/2019.

Choose a solution for connecting an on-premises network to Azure, 07/01/2018.

Compare storage options for use with Azure HDInsight clusters, 06/16/2019.

Container support in Azure Cognitive Services, 06/10/2019.

Continuous integration and delivery (CI/CD) in Azure Data Factory, 01/16/2019.

Create a bot with Azure Bot Service, 05/30/2019.

Create a new device template version, 03/25/2019.

Device connectivity in Azure IoT Central, 04/08/2019.

Device Simulation solution accelerator overview, 12/02/2018.

Feature comparison: Azure SQL Database versus SQL Server, 05/09/2019.

HDInsight 4.0 overview (Preview), 10/03/2018.

Information Bot Scenario, 12/12/2017.

Ingest data from Event Hub into Azure Data Explorer, 07/16/2019.

Internet of Things (IoT) Bot Scenario, 12/12/2017.

Introduction to Azure Data Lake Storage Gen2, 12/05/2018.

Introduction to the Azure IoT reference architecture, 12/03/2018.

Introduction to the Geo Artificial Intelligence Data Science Virtual Machine, 03/04/2018.

Machine Learning Anomaly Detection API, 06/04/2017.

Manage devices in your Azure IoT Central Application, 06/08/2019.

Monitor cluster performance (HDInsight), 05/28/2019.

OPC Twin architecture, 11/25/2018.

Order device connection events from Azure IoT Hub using Cosmos DB, 04/10/2019.

Overview of Azure Digital Twins, 05/30/2019.

PolyBase scale-out groups, 04/22/2019.

Predictive Maintenance solution accelerator overview, 03/07/2019.

Provision a Geo Artificial Intelligence Virtual Machine on Azure, 03/04/2018.

Quickstart: Create an Azure Cosmos account, container, and items with the Azure portal, 07/11/2019.

Quickstart: Create an Azure Data Explorer cluster and database, 03/14/2019.

Quickstart: Create an Azure Data Lake Storage Gen2 storage account, 12/05/2018.

Quickstart: Create Apache Hadoop cluster in Azure HDInsight using Azure portal, 06/11/2019.

Quickstart: Find available rooms by using Azure Digital Twins, 06/25/2019.

Quickstart: Query data in Azure Data Explorer Web UI, 07/03/2019.

Quickstart: Try a cloud-based solution to manage my industrial IoT devices, 07/03/2019.

Quickstart: Try a cloud-based solution to run a predictive maintenance analysis on my connected devices, 03/07/2019.

React to IoT Hub events by using Event Grid to trigger actions, 02/19/2019.

Reference – IoT Hub endpoints, 06/09/2019.

Remote Monitoring solution accelerator overview, 03/07/2019.

Security considerations for data movement in Azure Data Factory, 06/14/2018.

Set up a device template, 06/18/2019.

The Azure Blob Filesystem driver (ABFS): A dedicated Azure Storage driver for Hadoop, 12/05/2018.

Time series analysis in Azure Data Explorer, 04/06/2019.

Tutorial: Add a real device to your Azure IoT Central application, 04/22/2019.

Tutorial: Create a custom simulated device, 10/24/2018.

Tutorial: Stream data into Azure Databricks using Event Hubs, 06/20/2018.

Types of insights supported by Power BI, 12/05/2018.

Use Apache Kafka on HDInsight with Azure IoT Hub, 11/05/2018.

Use features in the Connected Factory solution accelerator dashboard, 07/09/2018.

Ways to share your work in Power BI, 06/06/2019.

Welcome to CosmosDB, 07/22/2019.

What are Azure Cognitive Services?, 04/18/2019.

What are IoT solution accelerators?, 03/08/2019.

What are the Apache Hadoop components and versions available with HDInsight?, 06/06/2019.

What is Azure Data Explorer?, 09/23/2018.

What is Azure Databricks?, 06/07/2019.

What is Azure Event Grid?, 05/24/2019.

What is Azure HDInsight and the Apache Hadoop technology stack, 01/27/2019.

What is Azure IoT Central?, 04/23/2019.

What is Azure IoT Edge, 04/16/2019.

What is Azure IoT Hub?, 07/03/2018.

What is Azure Machine Learning service?, 05/01/2019.

What is Azure Sphere?, 05/01/2019.

What is Azure SQL Database Service?, 04/07/2019.

What is Azure SQL Data Warehouse?, 05/29/2019.

What is Azure Stream Analytics?, 05/15/2019.

What is Azure Time Series Insights?, 04/25/2019.

What is Connected Factory IoT solution accelerator?, 06/09/2019.

What is industrial IoT (IIoT)?, 11/25/2018.

What is Power BI?, 05/29/2019.

What is Power Query?, 10/15/2018.

What is the Anomaly Detector API?, 03/25/2019.

Other Web Site Sources

IEC (International Electrotechnical Commission). https://www.iec.ch.

 IEEE (Institute of Electrical and Electronics Engineers). https://www.ieee.org.

 IETF (Internet Engineering Task Force). https://www.ietf.org.

 ISA (International Society of Automation). https://www.isa.org.

 ISO (International Organization for Standardization). https://www.iso.org.

 Microsoft Azure IoT Device Catalog. https://catalog.azureiotsolutions.com.

 OASIS (Advancing Open Standards for the Information Society). https://oasis-open.org.

 OMG (Object Management Group). https://www.omg.org.

OPC Foundation (OPC UA industrial interoperability standard). https://opcfoundation.org.

OSIsoft (PI system). https://www.osisoft.com.

PTC (ThingWorx platform). https://www.ptc.com.

USDA Agricultural Research Service (photos courtesy of, used in Cognitive Services Custom Vision example). https://www.ars.usda.gov/oc/images/image-gallery.

W3C (Worldwide Web Consortium). https://www.w3c.org.

Index

A

Advanced Message Queuing
Protocol (AMQP), 24, 55, 57
Agricultural Research Service (ARS), 105
ARPANET, 2
Azure Blob File System (ABFS), 35, 96
Azure Data Box, 145, 156
Azure Data Box Edge, 157
Azure Databricks, 91
  cluster creation, 92
  initial view, 93
  notebook view, 94
  workplace, 94
Azure Data Explorer, 151
Azure Data Factory (ADF), 37
  Azure Monitor, 151
  connectors, 150
  copy activity function, 150
  data connectors, 148
  ELT, 148
  linked services, 148, 149
  pipelines, 148
Azure Data Lake Storage (ADLS), 31, 96, 152
Azure HDInsight, 35, 36, 95
Azure IoT, non-Microsoft
components, 38, 39
Azure IoT certification service (AICS), 68
Azure IoT device catalog
  AICS, 68
  capabilities and properties, 70, 71
  certification test, 69
  industrial protocols, 72
  security levels, 69
  web site, 68
Azure IoT Edge, 32, 52
Azure IoT Hub, 32, 33, 61
Azure Machine Learning service
  experiment results, 104
  generating workspace, 101
  Notebook, 102
  Visual Studio, 102, 103
Azure Machine Learning Studio
  drag-and-drop interface, 99
  experiment, 100
Azure management and deployment
  governance plan, 41
  portal (see Azure Portal)
  resilience
    network, 50
    vs. on-premises, 49
    services, 47, 48
    storage, 49
  Security Center, 50, 51
  subscriptions and resource groups
    enterprise hierarchy, 42
    RBAC, 43
    resource manager, 42
Azure ML Service, 38, 103, 104
Azure portal, 54
  Azure Advisor, 45, 46
  Azure Cost Management, 46, 47

199

R. Stackowiak, *Azure Internet of Things Revealed*, https://doi.org/10.1007/978-1-4842-5470-7

Azure portal (*cont.*)
    dashboard layouts, 44
    dashboard view, 43
    monitor interface, 44, 45
Azure public cloud, 29
Azure resource manager, 42, 43
Azure Security Center, 50, 51, 78
Azure Sphere, 52, 53
Azure Sphere microcontroller units
    (MCUs), 52, 71
Azure Stream Analytics, 33, 88–89
Azure Virtual Network (VNet), 153

B

Bot Framework, 117, 118

C

Cloud computing
    backend platforms, 23
    IaaS, 22, 23
    on-premises deployment, 22
    PaaS, 22, 23
    SaaS, 22, 23
Cloud to device (C2D), 24
Cognitive services, 38, 104–107
Computer Vision Service
    image testing, 107
    image training, 105, 106
    visual features, 104
Connected Factory solution accelerator
    automated provisioning, 139
    Azure resource group, 140
    cloud dashboard, 138, 139
    demonstration dashboard, 141
    OPC UA interface, 138
    RBAC, 139

Consumer packaged goods (CPG), 7
Cosmos DB, 37, 97–99

D

Databricks, 34, 91–95
Data catalog
    defined, 158
    metadata, 158
    sensor data, 159
Design thinking
    agile sprint approach, 183, 184
    full-scale operation, 166
    ideation (*see* Ideation)
    observe and research
        business processes, 167
        external influencing
          factors, 169
        financial statements and
          presentations, 169
        industry and government
          sources, 169
        interviews, 168
        journey model, 168
        opportunities, 167
        PESTEL analysis, 169
        problem solution process, 168
        typical information, 167
    phases, 165, 166
    problem definition
        capturing and prioritization, 172
        deprioritize, 172
        IoT initiatives, 170
        notes, 171
        points of view, 170
        ROI, 171
    production-ready solution, 166
    prototype creation, 180, 181

prototypes (*see* Prototypes,
  implementation)
testing, 165, 181, 182
Device Simulation solution accelerator
  automated provisioning, 141, 142
  Azure resource group, 142, 143
  custom device simulations, 143
Digital twin, 33
Directly device to cloud (D2C), 24

E, F, G

Edge devices, 51, 52, 55
Edge sensor and device
  AMQP, 57
  Azure IoT Hub, 56
  data, 56
  IIoT, 58
  OPC UA, IIoT walls, 59, 60
  OPC UA servers, 58
  OSI and TCP/IP model, 57
  physical considerations, 56
  UPS, 56
Extraction, loading, and
        transform (ELT), 148

H

Hortonworks Data Platform (HDP), 35, 95
Hub–spoke topology, 156

I

Ideation
  alternative approaches, 179
  Automated Output Metrics
    Measurement, 176, 177
  brainstorming, 173, 175

definition, 173
facilitator, 173
ideas, 173
Post-it Notes, 173
primary goal, 175
pros and cons, 177
score, 179, 180
solution themes, 174
SWOT, 177–179
technical capabilities, 176
Industrial Internet Consortium (IIC), 15, 16
Industrial Internet of Things (IIoT), 58, 59
Infrastructure as a Service (IaaS), 22
Integration and data source
  ADF (*see* Azure Data Factory (ADF))
  Azure Data Catalog (*see* Data Catalog)
  data transfer, 156–158
  on-premises networks
    ExpressRoute, 155, 156
    hub–spoke topology, 156
    VNet, 153
    VPN gateway, 153, 154
    VPN site-to-site, 155
  query services
    Azure Data Explorer, 151
    KQL, 151
    PolyBase, 151
    PolyBase scale-out group, 152
Intelligent devices, 2
Internet of things (IoT)
  devices, 24
  evolution
    NoSQL databases, 3
    relational databases, 3
    sensors, 3
    thermostat, 2
    timeline, 3, 4
  ISA 99, 25

Internet of things (IoT) (*cont.*)
   messaging protocols, 24
   RMF, 25
   SAL, 25, 26
IoT architecture, components, 87
IoT-based business
   agribusiness, 5
   automotive, 5
   aviation, 6
   banks and financial trading, 9
   communications and media
      transmission, 6
   construction, 7
   CPG, 7
   education and research, 8
   environmental controls, 8, 9
   healthcare payers, 9, 10
   high tech and industrial
      manufacturing, 10
   insurance companies, 11
   law enforcement and emergency
      services, 11
   media content and entertainment, 11, 12
   oil and gas companies, 12
   pharmaceutical and medical device, 12
   retailers, 13
   transportation and logistics
      management, 13, 14
   utility companies, 14, 15
IoT Central
   administrators, 120
   application, creation, 121–123
   builders, 120
   Contoso sample application, 124, 125
   create device templates, 123
   device explorer, 124–126
   device template, 129

   export data, 129
   jobs, creation, 128
   Microsoft web site, 120, 121
   operators, 120
   refrigerator dashboard, 127, 128
IoT Edge runtime
   containers deployment
      Docker, 64
      IoT Edge Agent, 64
      key components, 64
      machine learning, 65
      SQL Database Edge, 65
   IoT Edge Agent, 61
   IoT Edge device
      identity translation gateway, 63
      protocol translation gateway, 62, 63
      transparent gateway, 62
   IoT Edge Hub, 61
   security framework
      attestation, 67
      authentication, 66
      authorization, 67
      device manufacturers, 66
      ISO/IEC 11889, 67
      principles, 65
      Security Manager, 66
      SGX, 66
IoT Hub Device Provisioning Service
   allocation policy, 84
   automated steps, 84
   availability, 84, 85
   features, 83
   tasks, 83
IoT Hub service
   activity, 79
   cloud-based services, 75
   configuring and controlling, 74

event routing options, 80, 81

managing, 78

message routing options, 80

monitor performance, 81

multi-model simulation, 82

option levels, 77

resource assignment and naming, 76

simulation accelerator, 82

tier and scale selection, 77

IoT reference architectures, 30

domains, 15, 16

IIC, 15

ISA-95, 15

in IT architecture (*see* IT architecture)

Open Software Foundation, 17

IoT SaaS solution

accelerators, 40

Microsoft Dynamics 365, 41

Microsoft PowerApps, 41

IoT solution accelerators

home page, 130, 131

MVC architecture, 130

web site, 130

IT architecture

Lambda architecture, 18

bidirectional exchange, 21

conceptual view, 19

IoT components, 19, 20

traditional batch-oriented

infrastructure, 17, 18

J

Jupyter Notebooks, 38, 102

K

Kusto Query Language (KQL), 151

L

Low-latency analytical

processing (LLAP), 35, 95

M

Manufacturing execution

system (MES), 58

Message Queue Telemetry

Transport (MQTT) protocol, 32,

55, 57

Microsoft Azure PaaS

APIs, 38

Azure Data Lake Storage, 35

Azure IoT Hub, 32, 33

components, 31

Cosmos DB, 37

Databricks, 34

data stores, 37

digital twin, 33

HDInsight

HDP, 35

open-source components, 35, 36

programming languages, 36

IoT architecture, 31

Stream Analytics, 33

time series, 34

tools, 37

Microsoft documentation, 194–196

Microsoft SQL Data

Warehouse, 146

N

National Institute of Standards and

Technology (NIST), 25

O

Office 365, 29
Open Platform Communications (OPC), 32
OSIsoft
  PI Server components, 160
  PI System, 160
  PI system integration, 160, 161

P, Q

Platform as a Service (PaaS), 22, 29, 129
Power BI, 33, 37
  dashboard, 114
  dashboard Q&A, 114
  datasets/dashboard tiles, 115
  layout of tables, 109
  mobile view, 112
  new report, 113
  output, 116
  reports and dashboards, 117
  starting, 108
  typical report, 110
  web browser view, 111
Predictive maintenance
  automated provisioning, 134, 135
  Azure resource group, 135, 136
  demonstration model, 136
  ML Studio Workspace, 137
  RUL dashboard, 137, 138
Preexisting footprints
  legacy data, 146
  SQL Database, 146
  transactional systems, 146, 147
Prototypes, implementation
  implementation strategy, 189, 190
  operational considerations, 187–189
  roadmap, 190, 191

ROI
  business value, 186
  formula, 185
  illustration, 186, 187
  TCO components, 185
PTC, ThingWorx, 161, 162

R

Recovery point objectives (RPOs), 85
Recovery time objective (RTO), 85
Remaining useful life (RUL), 137
Remote Monitoring solution accelerator
  automated provisioning, 132
  Azure resource group, 133
  configuration, 132
  dashboard, 133, 134
Return on investment (ROI), 171
Role-based access
    control (RBAC), 43, 139

S

Security assurance levels (SALs), 25
Security Information and Event
    Management (SIEM) tool, 67
Semi-structured data
  Cosmos DB, 97
    data consistency levels, 97, 98
    initial configuration, 98
    loading data, 99
  HDInsight, 95
    three-step process, 96, 97
Service Level Agreement (SLA), 84
Skills, IoT-based solution
  design thinking, 27
  key areas, 26, 27
Software as a Service (SaaS), 22

Software Guard Extension (SGX), 66
Solution accelerators, 54, 119
Spatial intelligence graphs, 33

T

Time Series Insights (TSI), 34
    creation summary, 90
    explorer, 91
    metadata, 89
Total cost of ownership (TCO), 185

U

Uninterrupted power supply (UPS), 56

V

Visual Studio, 38, 102

W, X, Y, Z

Web Site sources, 196, 197
Windows 10 IoT, 31, 53